Exploring the Attitudes of Students Aged 13-16 Towards Mathematics

Majed Saeed Alharthi

Title: Exploring the Attitudes of Students Aged 13-16 Towards Mathematics

ISBN: 979-8-89248-557-9

Author: Majed Saeed Alharthi

Cover image: www.pixabay.com

Publisher: Generis Publishing
Online orders: www.generis-publishing.com
Contact email: info@generis-publishing.com

Abstract

Many students in Saudi Arabia complain that they achieve poor results in mathematics and appear to have a negative attitude to the subject. The current study therefore investigates students' attitudes towards mathematics, using a mixed method approach, in 13-16 years old students from the intermediate level of education. The age was important because this period is thought to be crucial to the formation of lasting attitudes and opinions. The first main research question, the study aims to address is therefore, "What are students' attitudes towards mathematics in Saudi middle schools?" The study also aims explore gender issues using a second research question, "Is there any significant gender difference in students' attitudes towards mathematics?"

The survey was conducted in two government-run middle schools in Saudi Arabia, one boys' school and one girls' school, because Saudi Arabia has an exclusively single-sex education system. 180 participants (90 boys and 90 girls) were asked to complete a questionnaire to show their views about mathematics. Qualitative and quantitative data were collected at the same time, but the analysis of each type of information was conducted separately to enable the full picture to be understood. T-test was used to explore any gender differences.

The main findings were that both male and female students in two Saudi middle schools show a positive attitude towards mathematics, with female students being slightly more positive than males. The factors which affect attitudes towards mathematics in both genders include the usefulness of mathematics in everyday and future life and career, the teacher, enjoyment of mathematic and the difficulty of the subjects. Girls' attitudes towards mathematics were also affected by the influence of social media personalities and internet teachers.

Acknowledgments

First and forever, all praise to Allah, the almighty for allowing me to complete this research and write this dissertation. I would like to express my heartfelt gratitude to my supervisor, Dr Irene Biza, and professor Elena Nardi, for their professional guidance and support during all stages of this study. I am also very grateful to all those who completed the questionnaires and the open-ended questions, which were an integral part of this project. I would like also to extend my sincere thanks and appreciation to my family and friends, in Saudi Arabia, for their continual support and encouragement throughout my postgraduate study in the UK.

Majed Alharthi

Dedication

To my dear parents, wife, daughters, brothers, and sisters

with sincere love and respect...

Table of Contents

Abstract...

Acknowledgments i

Dedication ii

Chapter1: Introduction.. 2

1. Research Introduction.. 2

   1.1 Research Background and Problem.. 2

   1.2 Research Aim and Objectives... 2

   1.3 Significance of the Study: 3

   1.4 Dissertation Structure 4

Chapter 2: Literature Review 7

2. Related Literature 7

   2.1 Definition of Attitude Toward Mathematics 7

      2.2.2 Social Factors 13

   2.3 Attitude towards mathematics 14

   2.4 Conclusion 15

Chapter 3: Research Methodology 17

3. Research Methodology 17

   3.1 Research Design 17

   3.2 The Education System in Saudi Arabia... 17

   3.3 Research Method and Technique 18

   3.4 Development of the Questionnaire 20

   3.5 Targeted Population and Sampling Technique.. 25

   3.6 Data Collection Procedure 26

   3.7 Data Analysis Process 27

   3.8 Reliability and Validity of Attitude Measures.. 28

   3.9 Ethical consideration 29

   3.10 Problems and Limitations... 29

Chapter 4: Data analysis and Discussion.. 32

4. Data Analysis and Discussion 32

   4.1 Introduction 32

   4.2 The reliability analysis 33

   4.3 Result of Mathematics Anxiety 35

   4.4 Result of Confidence with Mathematics 38

   4.5. Result of Perceived Usefulness of Mathematics 42

4.6. Result of Enjoyment of Mathematics.. 45

4.7. Results of mathematics class environment relationship with teacher and peers........... 47

4.8 Overall result of students' attitudes towards mathematics.. 49

4.9 Analysis the open-ended questions. 51

    Usefulness of mathematics in everyday life: 52

    The importance of the teacher: 53

    Mathematics for career and 'future life':... 53

    Enjoyment of mathematics: 54

    The Notable Difficulty of mathematics: 54

    Teachers on the internet and social media personalities:... 55

    The links between mathematics and the Islamic religion, and mathematicians: 56

    The encouragement and support received from parents: 56

    Mathematics is an arithmetic subject which help us in daily life, science, and career:.... 57

    Mathematics is addition, subtraction, multiplication, division, shapes, geometry, rules, equations: 58

    The summary of the two previous questions: 58

Chapter 5: Conclusion.. 61

5. Conclusion 61

    5.1 Conclusion 61

    5.2 General and Academic Contributions 63

    5.3 Personal Outcomes.. 64

    5.4 Recommendations and Limitations... 65

References:... 67

Appendices:.. 80

Chapter 1
Introduction

Overview: This chapter provides an introduction for the reader to the research topic. Firstly, it presents a brief background to this area of study and then addresses the aims and objectives of the research, along with the research questions and an explanation of the significance of the study. Finally, this section outlines the contents of each chapter, and shows the methodological approach to the research.

Chapter1: Introduction

1. Research Introduction

1.1 Research Background and Problem

Since I started teaching in 2008 in middle school in Saudi Arabia, many students, at various levels, have initially expressed to me a negative attitude toward mathematics which they cannot explain. In addition, many are ignorant of the vital role of mathematics in our daily lives. Mathematics is the cradle of all creation, and everyone, needs mathematics in their day-to-day life. Therefore, mathematics is considered one of the most essential core subjects in a school curriculum (Orton et al., 2004).

According to Third International Mathematics and Science Study (TIMSS), the mathematics performance of Saudi students was very low in 2015 compared to 2011, with a sharp decline in their mathematics ranking for middle schools. Out of 39 countries they ranked lowest, despite Saudi Arabia having applied a new mathematics curriculum since 2012, which is similar to curricula published by McGraw-Hill (Ministry of Education in Saudi Arabia). These alarming results have encouraged me to focus my attention on investigating the reasons for Saudi Arabia's poor performance in this field. A number of studies have been conducted to ascertain the factors influencing students' performance in mathematics, a major factor being students' attitude toward mathematics (Mclead, 1992). In addition, many mathematics educators believe that, "If research on learning an instruction is to maximize its impact on students and teachers, affective issues need to occupy a more central position in the mind of researchers" (Mcleod, 1992, p.575). Studies have also analyzed, whether negative or positive attitudes toward mathematics, will affect academic success (Pobham, 2005; Beaton et al., 1996). Furthermore, there is real concern among many Saudi educators and teachers that this serious issue will have a negative impact on the future of mathematics in their country.

1.2 Research Aim and Objectives

Several studies have shown that the early years of adolescence is considered a critical period for the development of attitudes toward mathematics (see, for example, Johnstone and Hadden, 1983). Therefore, this research focuses on students in Saudi middle schools (aged 13-16). It identifies the essential factors behind attitudes toward mathematics, and examines what affects them from the student perspective. The first main research question the study aims to address is, "What are students' attitudes towards mathematics in Saudi middle schools?" Based on first main research question, there are two subsidiary research questions:

a. What factors affect students' attitudes about learning of mathematics?

b. What are students' attitudes about "what mathematics is"?

Furthermore, this research focuses on both boys and girls from Saudi middle schools. Although, the Ministry of Education in Saudi Arabia follows a policy of gender-based separation at all levels (Sait et al.,2004), the mathematics curriculum is the same for girls and boys (Ministry of Education, 2006). I explore the issues of gender, as summed up by second main research question, "Is there any significant gender difference in students' attitudes towards mathematics?" The following steps will contribute to achieving the purpose of the study:

• Conducting a thorough review of the related literature regarding understanding factors that affect students and create a negative attitude towards mathematics. It will also examine the steps which can be taken to minimize this negative attitude and discuss how educators can foster positive attitudes toward mathematics.

• A questionnaire will be used to collect data, based on my reading of the literature.

• An objective method of data collection and analysis will be used to ensure that the study has both validity and reliability.

• A list of recommendations will be prepared for the attention of decision-makers in Saudi Arabia.

1.3 Significance of the Study:

This study addresses the requirement to measure students' attitude toward mathematics, an issue which risks neglect in Saudi education. It will also raise awareness of how important attitudes are important in building pupils' success. By describing the relationship between pupils and mathematics, the study will help educators enhance students' mathematics performance. Furthermore, the study findings will benefit students by enabling them to improve their attitude toward mathematics. In addition, this study is important for several further reasons. Firstly, it will aid future, beneficial, educational developments in Saudi

Arabia, especially by underlining the essential nature of mathematics as a foundation for many other subjects. Secondly, this study might contribute towards achieving the Saudi Vision 2030, in which the most important element is educational development (Vision2030, 2018). Finally, the study could help the Ministry of Education in Saudi Arabia to prepare capable, qualified mathematics teachers who can increase positive attitudes towards mathematics in their students because, without positive attitudes, a student may not be stimulated to learn (Jung, 2005).

1.4 Dissertation Structure

This dissertation is divided into five sections, as follows:

Research Introduction: This chapter presents an introduction and background information to the research topic and problem, a statement of the research aims, objectives and questions, the significance of the study.

Literature Review: This chapter provides an overview of attitudes related to mathematics and a description of the importance of attitudes and different definitions. It examines factors affecting the students' attitudes towards mathematics in middle schools, gender differences in those factors. It concludes with a discussion of findings to date.

Research Methodology: This chapter presents an overview of the research methodology used to conduct this research and justifies each relevant methodology. The chapter also provides an overview of the data collection techniques used in the research and the strategies employed to analyse the data.

Results Analysis: This chapter analyses the data collected, presents the results obtained and discusses the findings.

Conclusions and Recommendations: This chapter draws conclusions from this research, answers the research questions, discusses how and to what degree the aims and objectives of this study has been accomplished. It then summarises the limitations of the study and provides recommendations for future research.

Chapter 2

Literature Review

6

Overview: Future learning is affected by students' attitude towards mathematics. How students experience their mathematics learning will shape their attitude towards the subject. This chapter, therefore, focuses on a review of literature that addresses attitudes related to mathematics, the factors affecting these attitudes and gender differences of each factor.

Chapter 2: Literature Review

2. Related Literature

2.1 Definition of Attitude Toward Mathematics

Attitude is a major factor in the learning and teaching process in mathematics (McLeod, 1992), helping us to understand ourselves, the world around us and our relationships (Reid, 2006). Several studies have shown that positive or negative attitudes affect the performance of students in mathematics (Hungerman, 1967; Hart, 1989). Attitude can be defined as "The affect for or against the psychological object" (Thurstone, 1931, p.261). However, this definition focuses exclusively on affect and ignores other aspects of human experience. In 1935, Allport developed this definition to cover "mental and neural state of readiness to respond, organized through experience, exerting a directive and or / dynamic influence on behaviour" (p.810). This definition focuses on behaviour and neglects the cognitive and affective factors. In 1992, Oppenheim combined and summarized the above definitions to explain attitude in an acceptable way for most researchers "Attitude is a state of readiness, a tendency to respond in certain stimuli …… attitudes are reinforced by beliefs, often attract strong feelings which may lead to particular behavioral intents" (Oppenheim, 1992, p.174-175; Ramsden, 1998). This definition includes three components: cognitive (beliefs, thoughts, attributes), affective (feeling, emotions) and behavioural information (past event, experiences) (Maio et al., 2010). These components all impact on student attitudes (Eagly and Chaiken, 1993). Oppenheim (1992)'s definition concurs to some extent with Hart's multidimensional definition (1989) which includes three components of attitude: emotional response, beliefs regarding the subject, behaviour related to the subject. In contrast, McLeod (1992) and Haladyna et al (1983) claim that attitude to mathematics is just an emotional disposition, either positive or negative, towards the subject. Despite attempts to define attitude toward mathematics, a number of researchers (e.g. Di Martino and Zan, 2011; Hannula, 2002) show attitude to be an ambiguous construct. Attitude is often ill-defined and requires better theoretical development. Di Martino and Zan (2011) demonstrate that research on attitude falls between a large number of different academic disciplines (e.g. mathematics, psychology, cognitive science, anthropology). The present research position follows McLeod (1992)'s multidimensional definition that includes various types of feelings towards Mathematics, such as love, hate, anxiety, interest, and a perception of the usefulness of Mathematics in life. Thus, statements such as "I like Mathematics" or "I'm interested in Mathematics" are defined as attitudes.

2.2 Factors Affecting Students' Attitude toward Mathematics

The International Association for the Evaluation of Educational Achievement (IEA) in 1996 conducted The Third International Mathematics and Science Study (TIMSS) on middle schools. Beaton et al. (1996) considered this one of the largest and most ambitious studies and noted that a positive connection between achievement in mathematics and students' attitudes towards the subject was often observable.

Saudi middle school students' attitudes towards mathematics might be expected to be negative, given their TIMSS results, with potentially far-reaching consequences. Negative attitudes stay with students at all stages of their education and may result in them later opting out of mathematics (Aiken, 1969), thus excluding themselves from many areas of study and potential future careers.

A range of interrelated factors that influence students' attitudes can be identified and divided into two groups. Firstly, there are personal factors related to the students themselves (e.g. mathematical anxiety, confidence with mathematics, achievement in mathematics, the students' ignorance of the importance of mathematics) and secondly, social factors involving society, such as teachers, mathematics class environment and peers.

2.2.1 Personal Factors

> **Mathematics Anxiety**

Dreger and Aiken (1957)'s concept of "number anxiety" has been an important research area ever since its introduction and is of increasing interest. Mathematics anxiety measures tend to associate quite closely with attitude measures, where mathematics anxiety is considered one of the factors which influence students' attitude toward mathematics and vice versa (Tahar et al., 2010; Dowker et al., 2016, Belbase, S., 2010). Furthermore, emotional and motivational factors, including belief and anxiety, can help us understand mathematics achievement (Soleymani and Rekabdar, 2016; Namkung et al., 2019). Richardoson and Suinn (1972, p.551) define mathematics anxiety as "… a feeling of tension and anxiety that interfere with the manipulation of numbers and the solving of mathematics problems in wide variety of ordinary life and academic situation". Liebert and Morris (1967) see that there are two separate dimensions of anxiety. The first, worry, is cognitive, and concerns performance and the effects of failing. The second is the affective aspect or "emotionality". This involves the autonomic reactions connected to nervousness and tension in testing situations. This concurs with Wigfield and Meece (1988)'s findings about mathematics anxiety in older children and adolescents, which exemplifies mathematics anxiety as a construct.

There has been extensive research into the importance of gender in this field. Many studies reported that females in intermediate stages in schools were significantly more anxious than males (Meece el at., 1990; Wigfield and Meece, 1988; Bonnstetter, 2007). Moreover, in a study of 433 British 11 to 15-year olds, Devine et al. (2012) found no significant performance differences between boys and girls who completed a mathematics test and a questionnaire about mathematics and test anxiety. However, girls were significantly more anxious in mathematics and tests than males. In contrast, Olmez and Ozel (2012) reported that sixth and seventh grade male students were significantly more anxious than females, although the weight of research evidence suggests that girls have greater mathematics anxiety (Dowker et al, 2016). Numerous studies suggest that mathematics anxiety starts around age eight, reaches a peak in middle school and continues to increase with age (Scarpello, 2007). The "boys are better than girls" stereotype, might increase mathematics anxiety in females (Dowker et al 2016, Wigfield and Meece, 1988). However, for cultural reasons, Saudi girls might not suffer from this stereotype.

Based on the above literature, it is possible that differences between boys and girls regarding mathematics anxiety could be gender-based or involve other gender-related factors such as cultural, environmental or educational context (developed vs. developing country). For example, Reilly et al. (2019), reporting results from the 2011 (TIMSS), found that boys reported more favourable attitudes towards mathematics than females in most countries, while some countries in Middle East showed a reversal of this trend (e.g., Oman). Indeed, the variability of gender differences in STEM across cultures suggests that environmental and cultural factors are significant in the development or suppression of these differences (Reilly et al, 2019). In addition, distinctive types of religion, language and mathematics itself are generated by particular cultural groups (Bishop, 1988). Therefore, culture appears to play a role in gender differences in terms of attitude toward mathematics. The present study (in 13-16-

year olds) may contribute to understanding the effects of gender differences in this field, in Saudi Arabia.

➢ **Confidence with mathematics**

Confidence in mathematics is a belief (Fishbein and Azjen, 1975). If students believe that they can do mathematics, they will have confidence in mathematics. The present study takes adopts the definition that includes the concept of confidence in mathematics in general, not in particular mathematics topics. Thus, statements such as "I am confident in solving mathematics problems" or "I find mathematics easy" are defined as confidence in mathematics in general.

Low confidence in mathematics was amongst the factors affecting students' attitudes towards mathematics and achievement (Zimmermann, 2000; Çiftçi and Yıldız, 2019). According to Gniewosz et al (2012), transition into adolescence negatively affects confidence and performance in mathematics, particularly when students change schools. I also noted that some of my students claimed that their confidence was high in elementary school. This concurs with Bandura (1986,1997)'s observations that young students tend to overestimate their own abilities. It is, therefore, important for teachers to protect children from the disappointment of failures, especially during the transition into adolescence. Moreover, teachers can build confidence by encouraging students to believe that they can "do" mathematics. If teachers emphasise the importance of effort over innate ability, this will help instil a positive attitude to mathematics.

Confidence has been shown to have a possible gender aspect. Several studies suggest that boys and girls are equally confident in their mathematics ability in elementary school but that boys become more confident than girls in middle and high school (Pajares and Graham,1999). These results are linked to the previously-mentioned transition into adolescence and changes in confidence levels (Gniewosz et al., 2012; Bandura, 1986,1997). Moreover, this gender difference in mathematics confidence, sometimes called the ''confidence gap'' (Sadker, M. and Sadker, D, 1994), appears to begin in middle school (Fennema and Hart,1994). One reason for this, highlighted by Recber et al. (2018) is that, as in the case of mathematics anxiety, this may reflect stereotypes. Furthermore, girls' low confidence in mathematics could also be associated with anxiety (Parsons et al., 2009). However, Pajares and Graham (1999) found that there was no significant difference in confidence of male and female students towards mathematics at middle school.

This study will explore further the importance of confidence in attitudes to mathematics in Saudi middle schools (13-16), including the significance of gender differences.

> ➤ **Students' perceptions on the learning of mathematics and the usefulness of mathematics.**

Students' perceptions on the learning of mathematics originate from past experiences, derived from parents, teachers or even the media. These instil bad or good perceptions about mathematics, which in turn affect their attitude toward mathematics (Pepin, 2011). Morrisett and Vinsonhaler (1965) claim that childhood experiences may create attitudes toward mathematics, which endure throughout students' academic careers (Aiken,1969), and affect them during its intermediate stage (Syyeda, 2016). Students have varying perceptions about mathematics in the intermediate stages. For example, Pepin (2011) found that in a comparison between English and Norwegian 11-16 year olds, seven themes emerged related to their perception of what mathematics is, why they liked, or disliked mathematics, and their perception of being/not being able to be successful in mathematics, including later life, interest at challenge, group work, the teacher's role, family support and examinations.

Some of the above points such as "Mathematics for jobs and 'later life' and "the role of the teacher" are related to Dobie (2019)'s findings concerning the perspectives of middle school students about the usefulness of mathematics.

Adelson and McCoach (2011, p.226) defined the perceived usefulness of mathematics as "a person's beliefs about the practical use and applicability of mathematics currently and in relationship to his or her future". In Saudi Arabia, I often heard my students asking, "How is this useful?" and, "When are we going to use this?". If students realize the usefulness of mathematics, they will be better motivated to study, learn and practice it (Pajares and Miller, 1994). Moreover, students' perceptions of the usefulness of mathematics is strongly correlated with their participation and achievement in mathematics (Chipman, 2005). However, I am concerned about the widespread beliefs or mathematical myths that "learning mathematics is a question more of ability than effort" (McLeod, 1992, p.575). Therefore, some of students believe that mathematics only suits those with a special talent for it. This creates a bad image of mathematics that remains in their minds and may endure for generations, affecting perceptions about its usefulness.

Both old and recent studies indicate that girls tend to hold more negative perspectives toward mathematics in middle schools. For example, Ding et al (2015) studied the attitudes of 4,236 students across grades 6 to 9 and found that 85.8 % of 2,153 boys expressed a liking for mathematics, as opposed to only 77.5 % of 1,703 girls. Moreover, in examining affective or attitudinal variables in middle school students, Fennema-Sherman (1978) found that more boys than girls believed that mathematics would be useful to them. However, Uwineza et al. (2018) found that in 15 to 18-year olds, both genders shared perceptions about the importance of the subject, although boys had negative perceptions about girls' abilities to succeed in mathematics.

Negative perceptions among students who do not like mathematics or find it useless might be detrimental to the image of mathematics in Saudi Arabia. It is therefore important for this study to investigate Saudi students' perceptions on the learning and usefulness of mathematics in middle schools, since these are a major factor affecting students' attitude towards mathematics.

> **Enjoyment of Mathematics**

Adelson and McCoach (2011, p.226) defined enjoyment of mathematics as "the degree to which a person takes pleasure in doing and learning mathematics". Enjoyment is one of the main constructs that measure mathematics attitude for middle school students (Aiken, 1976). According to PISA (2012) enjoyment and interest foster mathematics learning (OECD, 2013). Therefore, if enjoyment increases, attitude will improve. Moreover, enjoyment of and interest in mathematics give intrinsic motivation, where intrinsic motivation represents a principal source of enjoyment (Ryan and Deci, 2000). Thus, if you enjoy mathematics and have an interest in it, you will study it for its own sake rather than to pass a mathematics exam.

In terms of gender and interest, Leder and Forgasz (2002) found that of 800 middle school students, girls demonstrated more interested in mathematics and enjoyed it more, while boys tended to find it hard and boring. However, in an even larger study, according to PISA (2012) results published by OECD (2013), male students were significantly more interested in mathematics and enjoyed it more than girls. Høgheim and Reber (2019) replicate these results, showing that male students report higher interest in mathematics than do female students in middle school.

Finally, males usually consider mathematics to be more enjoyable and to have more meaning than their female counterparts (Høgheim and Reber, 2019) This study will investigate this aspect of attitude towards mathematics in Saudi middle schools and any significant gender difference in enjoyment of mathematics.

2.2.2 Social Factors

Mathematics class environment relationship with teacher and peers.

Attitudes toward mathematics can result from social factors such as teachers, mathematics class environment and peers (Yang, 2013). In terms of teachers, Ruffell et al. (1998, p.1) suggested that the, "Teacher's attitudes to mathematics is increasingly put forward as a dominant factor in children's attitudes to mathematics." Moreover, there are many studies which have indicated that student attitudes toward study, teachers, methods, and the overall school climate are influenced by the teacher-student relationship (Torrance et al., 1966 ; Fraser and Fisher, 1982; Hartmut, 1978). In middle schools, certain factors influence this relationship. For example, Roeser et al. (1998) observed that perceived unfair and disrespectful treatment by teachers undermined the academic motivation of adolescents and may exacerbate emotional upset and anger, with lasting effects (Aiken,1969). Aiken and Dreger (1961) blamed teachers for making college students hate mathematics. However, it is unfair to describe teachers as creators of negative student attitudes toward mathematics, since these can be the result of poor student performance in mathematics, the nature of mathematics itself, affecting student images of the subject, or other factors (Aiken,1969; Orton and Wain, 1994).

In terms of learning environments, support from teachers whether academically or affectively is one of the strongest factors affecting the mathematics classrooms in middle schools (Sakiz et al., 2012; Newman and Schwager, 1993). Teachers seek to foster a classroom atmosphere which is conducive to developing a positive attitude, which influences whether or not mathematics learning flourishes (Orton and Wain, 1994). Haladyna et al. (1983) also stress the impact of the teaching process and the classroom environment on students' attitudes towards mathematics. Learning environments are social, however, and elementary school children, especially girls, are influenced by the attitudes of their peers (Shapiro, 1961). Goodenow (1993a) noted that students experience a sense of belonging when they connect positively with their classmates. This is one of the most important factors in motivation, dedication and engagement (Goodenow, 1993a; Osterman, 2000), which in turn affects their attitude towards mathematics, especially as peers play an important role in development of mathematics attitudes during adolescence (e.g., Berndt 1979). However, in middle school, it is more difficult for students to establish meaningful relationships with their classmates than in elementary school (Hicks, 1997) . This may weaken the sense of belonging among students in middle school and affect their attitude.

In terms of gender, Asante (2012) found that for students aged 16 to 21 years, teacher attitudes, the school environment and beliefs about mathematics contribute to the gender differences in attitudes towards mathematics. Campbell (1995) points out that "teachers are part of the cause of the differential gender differences that exist. . . they have the power to contribute to helping to eliminate those practices" (p. 233). Thus, differences in female and male attitudes might reflect differential treatment by teachers. For example, in middle school, the teachers' stereotypes significantly affected the students' stereotypes which in turn will negatively affect girls' attitude toward mathematics (Keller, 2001). However, this occurrence could be weak in Saudi Arabia, due to gender-based separation at all levesl, which may help girls. For example, Lee and Lockheed (1990), found that girls had more positive attitudes to mathematics if taught by women teachers. However, in middle school, the gender of teachers had no impact on students' grading behaviours in some studies (e.g. Wiles,1992). Nevertheless, it might affect their attitude towards mathematics (Mallam, 1993).

Finally, when students are satisfied with the learning environment, teacher-peer relationships are more permanent and meaningful and will improve students' attitudes towards mathematics. This study will be applied in one boys' school and one girls' school in Saudi Arabia (aged 13-16), to explore social factors which affect student attitudes and investigate any significant gender differences.

2.3 Attitude towards mathematics

Mathematics appears to be more appealing to boys than to girls, as several studies have indicated that boys tend to have more positive attitudes than girls in middle school (Frost et al., 1994; Foxman et al., 1981; Else-Questet al, 2013). For example, Reilly et al. (2019), reporting results from the 2011 (TIMSS), found that boys reported more favourable attitudes towards mathematics in most countries. However, Reilly et al. (2019) found in Oman that more female students showed positive attitudes toward mathematics than males. One explanation could be due to   differences   in cultural environments and educational contexts between countries (e.g. Cogan and Schmidt, 1999). Regardless of the type of country, the gender-role socialization pattern in culture affects attitudes towards mathematics. For example, in their study of the effect, on attitude to mathematics of ethnicity and gender on Arab and Jewish eighth-grade students in Israel, Birenbaum and Nasser (2006) found a more positive attitude to mathematics amongst the Arab students. Furthermore, girls had more achievement-enhancing approaches than boys and were more successful. Birenbaum and Nasser (2006) stated that one possible explanation given for this was that, in traditional Arab families, daughters have lower status than sons and are less nurtured, spending more of their free time indoors helping with housework, while boys played outside. Girls might therefore use their studies to avoid some of the chores. They might then be motivated by parental expectations of academic success. In the Jewish group, however, gender differences reflected those found in Western cultures.

Finally, this study seeks to examine the factors that influence attitudes toward mathematics in two Saudi middle schools and recommend strategies to improve attitudes.

2.4 Conclusion

This chapter has examined the nature and importance of attitudes towards mathematics and the influencing factors, most of which remain outside the control of teachers. However, teachers play a vital role in improving students' attitudes towards mathematics because when teachers improve their students' attitudes from negative to positive, their students will develop specific personal goals and motivation for mastering these topics. Therefore, they will work harder and persevere through challenges. In addition, students can understand mathematics deeply and remember the information long after their exams, due to the relationship between enjoyable, participatory learning and long-term memory (Willis, 2010). It has also been noted that there may be gender differences in attitudes towards mathematics and this may depend on cultural and environmental factors and gender role socialisation.

Chapter 3
Research Methodology

16

Overview: The following chapter provides an overview of the research methodology used in the current study and a justification of the research methodology selected. It discusses the research design, methods and techniques, the development of the questionnaire, the sampling techniques, data collection and analysis and how the validity and reliability of the data were guaranteed.

Chapter 3: Research Methodology

3. Research Methodology

3.1 Research Design

The question of whether to use qualitative and quantitative methods in educational research has been a matter of controversy, in the literature. Some researchers favour using only one or other of these methods. However, other commentators combine the approaches because this allows the strengths of each method to work in a complementary manner (Muijs, 2004; Cohen et.al, 2011 and Yin, 1984). Yin (1984, p.92) stated "…any finding or conclusion in a case study is likely to be much more convincing and accurate if it is based on several different sources of information…". In addition, researchers can have greater confidence in the validity of their results when they used a mixed approach because it allows for the collection of data which is both comprehensive and robust (Cohen et al., 2011). Therefore, in this study and, in line with Stromquist's (2000) vision that no one methodological instrument that can guarantee the complete truth, both qualitative and quantitative methods was adopted. Moreover, quantitative methods alone would not be enough to tackle my research questions. I need to find out, not only what students' attitudes to mathematics are, but also the factors influencing those attitudes. Qualitative methods are more appropriate to exploring the factors influencing students' attitude towards mathematics (Yılmaz et al., 2010). Finally, qualitative and quantitative data were collected simultaneously, but the analysis of each type of information was conducted separately to enable the full picture to be understood.

3.2 The Education System in Saudi Arabia

Since 1953, the Ministry of Education has been responsible for the Education system in Saudi Arabia. It consists of general education, higher education, special education, adult education and literacy (Ministry of Education, 2006). Compulsory schooling involves a 6-year elementary stage from age 6 to12, a 3-year intermediate stage and a 3-year secondary stage. Each academic year has two semesters, with the same subjects being taught in both. The semesters are between 18 and 20 weeks long, which includes time for examinations. The Ministry of Education in Saudi Arabia follow a policy of gender-based separation at all levels (Sait et al., 2004). The curriculum for mathematics is the same for girls and boys at all levels (Ministry of Education, 2006). Mathematics is a compulsory subject in elementary and intermediate and the first year of secondary school for both girls and boys. Once students reach the second or third years of secondary education, they do have a limited choice of subjects, choosing from three fields of study: Natural science, or Islamic and Arabic studies, or Administration science. More intensive mathematics studies are required of those who choose either natural science or administration science.

3.3 Research Method and Technique

Fishbein and Ajzen (1975) point out that attitude can be measured whether positive or not. However, Henerson (1987) claimed that attitudes are not as easy to measure as heart rate, because we can only infer it by considering what they say and do (Henerson, 1987). Generally, when researchers measure attitude, they should select the method which is suitable for the objective of their research and the research timeline.

In this study, I want to find out about students' attitudes toward mathematics in Saudi middle schools. There are direct approaches (e.g. questionnaires, interviews and projective pictures) and indirect approaches (e.g. observation, or deduction of attitudes from some aspect of behaviour) to attitude measurement (Lemon, 1973; Corcoran and Gibb, 1961). In terms of mathematics, direct approaches are appropriate for academic and applied research (Bohner and Wanke, 2002), but respondents who harbor a negative attitude towards mathematics may not wish to admit it to themselves. Thus, it cannot be guaranteed that direct approaches are completely valid. In fact, in most research, direct methods, especially questionnaires and interviews are used in measuring attitudes toward mathematics.

An interview involves at least two people meeting face to face, with the interviewer posing questions and the respondent answering them (Henerson et al., 1987). An interview could help the researcher to gain further insight into the student's attitudes regarding mathematics. However, interviewing students individually is time-consuming and there is a risk that the interviewer may unwittingly influence the respondents' answers (Henerson et al., 1987), especially with adolescent students. Thus, the questionnaire approach is more appropriate for young adolescents than interviews (Bill, 1973). Furthermore, interviewing female students would have been impossible for a male researcher due to strict gender separation laws. All Saudi schools are single sex establishments, and opposite sex visitors would be forbidden, even for the purposes of research. It would not have been helpful to my project to interview boys only, as I need to have the viewpoints of both genders. I therefore decided that it was more appropriate to use the survey method only. Furthermore, attitudes to mathematics are likely to be extremely complex and may be multi-dimensional, involving different themes, including teachers, confidence levels and anxiety. A questionnaire which poses a wide range of questions can provide a wealth of insights and detail about the factors influencing students' attitudes. As a result, I will select the questionnaire as the primary technique to collect data.

Oppenheim (1992, p.100) defines the questionnaire as, "not some sort of official form, nor is it a set of questions which has been casually jotted down without much thought. We should think of the questionnaire as an important instrument of research, a tool for data collection, the questionnaire has a job to do: its function is measurement". The questionnaire method is widely used for conducting research because it is undemanding in terms of cost and effort (Ahern, 2005), covering all aspects of a topic, and producing speedy results. Furthermore, researchers can inquire directly about people's attitude. In terms of respondents, this method could attract many students, even those who are shy, because respondents' identities remain confidential. In addition, respondents at school levels prefer using questionnaires (Oppenheim, 1992).

Nonetheless, the questionnaire has certain disadvantages. Firstly, respondents could misunderstand your questions which may result in wrong answers (Henerson et al., 1987). Secondly, Bryman and Bell (2018, p.234) said, " There is no opportunity to probe respondents to elaborate an answer". Finally, someone other than the intended person may complete the questionnaires (Walonick, 1993).

There are different kinds of questionnaires to measure attitude towards mathematics. For example, Thurstone, Likert and Osgood's Method are having been used in measurement of attitude towards mathematics by many researchers, and each of these methods has its special style (Thurstone, 1929; Thurstone and Chave, 1929; Heise,1970; Oppenheim, 1966, 1992). In this study, I will follow Likert-style of questionnaire. This type of questionnaire is frequently used in education research because it is simple to construct and to analyse and because it makes standardisation possible (Lowry and Turner, 2007). However, social desirability bias is known to be an issue with self-reported, Likert scale questionnaires. This is because participants tend to respond to the questions in ways they think others will approve of (Grimm, 2010). Ways of avoiding this problem include using indirect questions and not specifically mentioning the research's subject matter (Ipsos, 2013). This research questionnaire also uses two open-ended questions to give respondents an opportunity to express their own ideas and to communicate their feelings.

3.4 Development of the Questionnaire

In order to obtain an insight into students' attitudes mathematics, the questionnaire used in this study, was adapted from one used in previous studies by Fennema (1976) and Sherman, Ma McCoach (1997), Wigfield (1988) and Meece, and Sakiz et al. (2012), Wong (2012) and Chen, and Pepin (2011), and He (2007), and Mendoza (1996), Adelson (2011). My choices were validated with my supervisor, a questionnaire was developed and edited by deleting some questions and adding some statements, to focus on the specific research area, answer the research questions and fulfil the aims of the study. I discussed my questionnaire with experienced colleagues and experts in the mathematics curriculum in Saudi Arabia to ensure the questionnaire's validity and reliability. After that, the design of the questionnaire was established and approval was given by my supervisor. The questionnaire was translated into Arabic by experts to ensure it was of high quality.

Finally, questionnaire was distributed the to six Saudi students, aged 13-16, in Riyadh, Saudi Arabia to carry out a small-scale pilot study to ensure that the questions used lay language (Locke et al., 2013). This questionnaire was divided into three main parts. Part one involved two questions to obtain general information about participants (age of students, gender). The second part, using the Likert format as a measure (Likert, 1932), consisted of 29 statements which aimed to elicit information about what students' attitudes towards mathematics were, involving questions about the views of students on the following attitudes aspects: (Further information, refer to Chapter 2)

1- Mathematics anxiety

2- Confidence with mathematics

3- Students' perceptions on the learning of mathematics

4- The usefulness of mathematics.

5- Enjoyment of mathematics.

6- Mathematics class environment and relationship with teacher and peers.

Section three provided information about pupils' perceptions of why they liked, or disliked mathematics, and their vision of what mathematics is. The aims of these two types of question are to gain a lot of information about different aspects of attitudes toward mathematics. The questionnaire is shown overleaf.

What do you think about mathematics?

We want to find out what you think of mathematics. Please answer every question as honestly as you can! The questionnaire is anonymous, and your answer will not affect your school marks in any way. We hope that what you tell us will help those who are learning mathematics.

Please tick the box provided to show you would like to participate to this study ☐

Your views are very important ☺

Part one: about yourself (Tick one box)

1) Are you   ☐ Girl   ☐ Boy
2) Your age   ☐ 13-14   ☐ 14-15   ☐ 15-16

Part two: Think about mathematics

This part contains twenty-nine statements. You are asked to indicate the extent to which you think you possess the criteria mentioned in each statement regarding mathematics by putting ($\checkmark$) in the right square.

	Strongly disagree	Somewhat disagree	Neither agree nor disagree	Somewhat agree	Strongly agree
1. Mathematics teaches me to think clearly.					
2. I worry about how well I am doing in mathematics.					
3. Taking mathematics tests scares me.					
4. I am enjoying doing mathematics.					
5. My mathematics teacher encourages me at times when I don't do well in class.					
6. Mathematics is more difficult for me than for many of my classmates.					
7. Other students in my mathematics class take my ideas seriously.					
8. I'm confident I can understand the basic concepts taught in my mathematics class.					
9. I worry that other students might understand the problem better than me, when the teacher is showing the class how to do a problem.					

Statement					
10. Even thinking about my mathematics class makes me feel hopeless.					
11. Mathematics is related in my life.					
12. I worry about how well I am doing in mathematics more than other subjects.					
13. Mathematics is useful for my career.					
14. I think some people have a mind for mathematics and some do not.					
15. I think mathematics is useful only for tests.					
16. I find mathematics interesting and motivating.					
17. I enjoy participating in my mathematics class.					
18. I find mathematics boring.					
19. Mathematics helps me to understand reports and advertisements about prices, sale, percentages etc.					
20. I need a good memory to do mathematics.					
21. When the teacher says he/she is going to ask me some questions to find out how much I know about mathematics, I worry that I will do poorly.					
22. I am good at using mathematics to solve real-life problems.					

23. I find mathematics easy.					
24. Mathematics is an important subject for me.					
25. My teacher thinks I can do well in mathematics.					
26. My math teacher respects me.					
27. I am confident in solving mathematics problems.					
28. I think mathematics is useful in solving real world problems.					
29. I feel like my presence matters in my mathematics class.					

Part three: please answer the fowling questions

1- I like/ dislike mathematics, because ....
....
....
....
....
....

2- Mathematics is ....
....
....
....
....
....

Thank you for answering
All the best in your studies ☺

Figure 1: Questionnaire Form

3.5 Targeted Population and Sampling Technique

"The quality of a piece of research stands or falls not only by the appropriateness of methodology and instrumentation but also by the suitability of the sampling strategy that has been adopted" (Cohen et al., 2011, p.100). It is difficult to cover all the study population in a single study because of factors such as expense, time and inaccessibility to specific members of the targeted population. Therefore, researchers will then often need to be able to select a representative subset of the population. This subset is the sample (Cohen et al., 2011).

In this study, two Saudi "government-run middle-schools" were chosen from different areas in Riyadh city in Saudi Arabia. According to the Riyadh Education Directorate (2018), there are 1186 middle schools and more than 341,000 students. I selected the boys' school where I was a teacher two years ago and a girls' school where family members are teachers. Therefore, my sampling approach was a convenience sampling. As mentioned before, all Saudi schools are single sex establishments. This study was applied in one boys' school and one girls' school. By using questionnaires and through collaboration with female family members, I was able to access indirectly participants in both male and female schools. This gave a more rounded perspective on both male and female students' attitudes towards mathematics. In Saudi middle schools, there are three years groups (aged 13-16). Two classes from each level were selected randomly, one form each school. 90 male and 90 female students participated in this research. The table 3.1 shows the numbers of students who were involved in this study in relation to the level and gender.

	Girls	Boys	Total
Level 1 (age 13-14)	30	30	60
Level 2 (age 14-15)	30	30	60
Level 3 (age 15-16)	30	30	60
			180

Table 3.1 Samples used

3.6 Data Collection Procedure

The data collection work was carried out over a period of 10 days, during the Easter holiday 2019. Permission for conducting this study was obtained from the Ethics committee at UEA. The questionnaires were given to the headmasters of the schools. The headmasters of each school were requested to complete a form and sign a written agreement to indicate their consent for students at their schools and themselves to participate in the study (see Appendix A and B). A package of information that provided more details about the research was also distributed to the leaders of the schools, including the study information sheet, data collection calendar, ethical rules, and the consent forms for the parents (see Appendix C). Permission for parents and students was obtained. Then, headmasters of each school were asked to invite participants in the school to respond to the printed questionnaire. In terms of the questionnaire for the female school, it was distributed by a female family member from whom I was also obtained consent to assist me in my study.   This sample might address attitudes toward mathematics, which will help educators to realize how they can improve students' attitude toward mathematics. In addition, this quantitative data will enable researchers to generalise results, (Cohen et al., 2011), taking account of the study limitations. Resources were insufficient to study all Saudi middle schools. However, the sample size of 180 students was large enough to generalise to reflect the attitudes of the overall population of mathematics students in Saudi middle schools.

3.7 Data Analysis Process

In this study, and based on first main research question, quantitative data was used, in the form of descriptive statistics resulting from the questionnaire. Qualitative data analysis was also used, in the form of thematic analysis resulting from the open-ended questions. While, based on the second main research question, an independent t-test was employed to assess whether there were differing average values for gender.

> **Quantitative data analysis**

The quantifiable part of the survey was analysed using quantitative methods. Descriptive statistics, such as mean and standard deviation, were used to understand the overall attitude. Such statistics seek to gather information on the characteristics of the research subjects and their activities without inferences or predictions. They simply inform researchers what has been found, in a variety of ways (Cohen et al., 2011).

The process and results of statistical tests were calculated and carried out using SPSS Statistics 25 software program and Excel software program. Excel was used to compile responses and they were imported by version 25 of SPSS. The rating scale was Likert and it was coded in SPSS. There were no missing values, which in turn will provide ease of analysis and increased reliability (Field, 2009). There were questions that contain the negative expression. Therefore, reverse coding was done for them. Normality testing was carried out to ensure that the data was normal. This is because, in parametric testing, there is an underlying assumption that the data is normal (Field, 2009). In other words, I used Kolmogorov–Smirnov test and Shapiro–Wilk test as the numerical means of assessing normality. To test for gender differences, an independent t-test was employed to assess whether there were differing average values for two groups (Field, 2009). For this reason, it was appropriate to use the independent t-test on the measurement related to gender for second main research question.

> **Qualitative data analysis**

Open-ended questions were used to encourage the participants give their own responses about some aspects of their attitude towards mathematics, rather than having to select from a few possibilities chosen by the researcher. Thus, insight into the participants' personal perceptions about their attitude towards mathematics was made possible by employing the open-ended questions.

A qualitative method was employed to analyse non-quantifiable data from the open-ended items (textual), which were analysed using thematic analysis because this is a flexible way of analysing such open-ended responses (Braun and Clarke, 2006). Students' responses to questions: "Do you like mathematics?", and "What is mathematics?" were given codes and arranged according to the themes which emerged. This was conducted manually in a step-by step manner. The first stage of the process involved reading the transcripts repeatedly and noting down the initial issues which emerged. The second stage involved coding the data and identifying the verbal expression which related to each code. Thirdly, the codes were sorted according to the emerging themes. A table was used to organise the themes which had been identified when all the extracts from the data had been assimilated. In the fourth stage, a review of the themes was undertaken to investigate how they related to the whole dataset. The themes were then placed in groups according to subject, in line with the research questions. Classification of the responses was undertaken with great care and patterns within the answers were sought where appropriate. This was particularly helpful where one respondent provided contradictory responses and extra care was taken with thematic classification.

3.8 Reliability and Validity of Attitude Measures

Any researcher must endeavour to maximise the validity and reliability of their data, which are both crucial to the measurement of attitudes. Validity can be defined as the degree to which a study measures what it is supposed to measure (Chaiken and Eagly, 1993). On the other hand, reliability reflects how far a technique provides scores or values which are consistent when attitude measurements are taken numerous times, (Chaiken and Eagly, 1993). The Cronbach Alpha coefficient is a method of determining the internal consistency of items in a scale. This is achieved by investigating average inter-item correlation. Calculation using the Cronbach's alpha coefficients shows whether or not there is correlation between items and whether it is necessary to modify or remove certain items. This process is applicable to attitude scales (Gardner, P.L., 1995). In this study, the scales were expertly reviewed and pilot studies conducted to ensure validity (see section of development of the questionnaire).

3.9 Ethical consideration

Ethical considerations are an essential part of research, without which the value of the research can be seriously undermined. Permission for this research was granted by the ethics approval committee at the University of East Anglia and approval to collect data in Riyadh Schools for this survey, was provided by the Ministry of Education in Saudi Arabia. Written approval was sought both from the leaders of the targeted schools and parents to approve their children's participation in the study. Participants were informed that I would ensure the anonymity of individuals and organisations participating in the research, and that I would safeguard participants' privacy (Bryman and Bell, 2007). In addition, the questionnaire was anonymous. Participants were informed that they could withdraw while they were responding the questionnaire, or refuse to fill in the questionnaire, but not after handing-in the questionnaire. In addition, they were assured that no one could access the data except my supervisor and myself. If participants experienced any problems in relation to any aspect of the research, they could contact me by e-mail. Participants were also informed that the data was for research purposes only. Thus, this study has been conducted with meticulous attention to detail, to ensure anonymity and confidentiality.

3.10 Problems and Limitations

During the research process some limitations and problems became apparent. Initially, it was challenging to recruit enough participants, because some parents did not approve their children's participation in the study. All of them were from the boys’ school. For this reason, the publication area of 'the Parents/Carer Consent Form' was increased in boys’ school. Finally, there was some confusion about age selection because some level one students, who were 14 ticked the level 2 box because two boxes included this age (See figures 2).

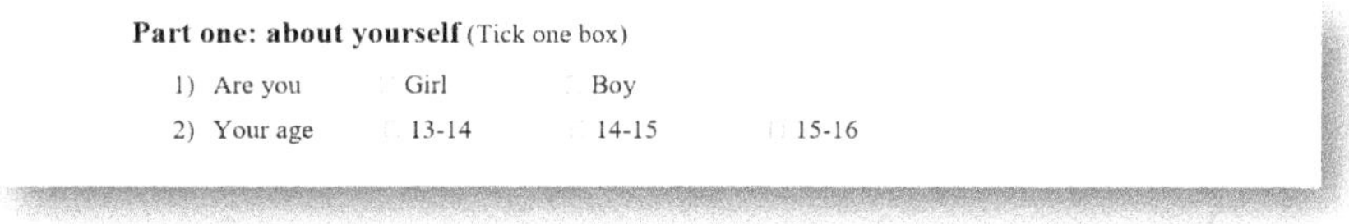

Part one: about yourself (Tick one box)

1) Are you Girl Boy

2) Your age 13-14 14-15 15-16

Figure 2: 3.1 Excerpt from the questionnaire for a question given in age data

However, this problem was solved by explaining the problem to the students and informing them that if their age was 14 and they were in level is 1, they should select first box (13-14), (See figures 3).

الجزء الأول: عن نفسك (ضع علامة في خانة واحدة)

الجنس: ☑ ذكر ☐ أنثى

العمر: ☑ ١٣ – ١٤ ☑ ١٤ - ١٥ ☐ ١٥ - ١٦

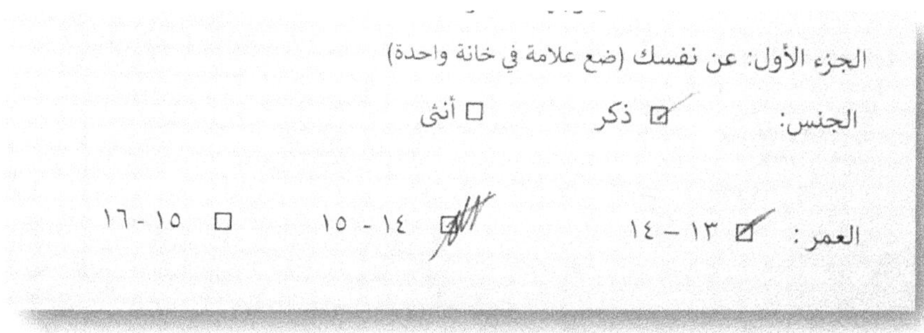

Figure 3: 3.2 Excerpt from the questionnaire for a question given in age data from one student.

Chapter 4
Data Analysis and Discussion

Overview: This chapter considers the analysis of the data collected and presents the result obtained. Each section contains a full discussion of those results.

Chapter 4: Data analysis and Discussion

4. Data Analysis and Discussion

4.1 Introduction

The questionnaire aimed to guage students' attitudes towards mathematics in two Saudi "government-run middle- schools", to identify the most significant factors driving those attitudes and to explore gender related differences. The students were asked to indicate their attitude towards mathematics by rating 29 statements on a 5-point scale, where Likert scale was coded in SPSS as 1 = "Strongly agree", 2 = "Agree", 3 = "Unsure", 4= "Disagree", and 5= "Strongly disagree". The responses for each part of each statement were summarised according to gender. The data were presented as frequency, percentage, mean, and standard deviation. Table 4.1 shows the mean values used as the basis for determining the agreement level. Statements were ranked according to standard deviation in those cases with equal mean values. Each respondent was given a final score determined according to the sum of the ratings for all statements.

A qualitative method was employed to analyse non-quantifiable data from the open-ended items. For each of the themes, an independent t-test was employed to assess differing average values for gender. This chapter discusses firstly what pupils' attitudes are towards mathematics and students' responses to the question of whether they liked or disliked mathematics and why, and the question, "What is mathematics?". Secondly, it considers the differences between male and female attitudes towards mathematics. (See Appendix D).

Mean values	Agreement level
1.00 – 1.79	Strongly agree
1.80 – 2.59	Agree
2.60 – 3.39	Unsure
3.40 – 4.19	Disagree
4.20 – 5.00	Strongly disagree

Table 4.1 The intervals have equal length.

4.2 The reliability analysis

Before analyzing the data of the study, it is essential firstly to assess the reliability of the questionnaire using Cronbach alpha (Cronbach, 1951), which is suitable for attitude scales (Dunn et al., 2014; Gardner, P.L., 1995). Cronbach's alpha reliability coefficient normally ranges between 0 and 1, George and Mallery (2003, p. 231) provide the following rules of thumb: "_ > .9 – Excellent, _ > .8 – Good, _ > .7 – Acceptable, _ > .6 – Questionable, _ > .5 – Poor, and _ < .5 – Unacceptable". The questionnaire of this study has six standards:

1- Mathematics anxiety.

2- Confidence with mathematics.

3- Students' perceptions on the learning of mathematics.

4- The usefulness of mathematics.

5- Enjoyment of mathematics.

6- Mathematics class environment and relationship with teacher and peers.

The Cronbach's alpha values for the six main attitude scales used in the questionnaire are shown in Table 4.2. The values of all scales, except the scale of students' perceptions on the learning of mathematics, range from 0.621 to 0.804, which satisfy the "Acceptable" criterion of reliability and provide evidence for the reliability of the questionnaire. The scale for students' perceptions on the learning of mathematics, rated "unacceptable" according to the criterion. Therefore, I combined it with analysis of the open-ended questions, with which this scale fitted in perfectly.

The overall Cronbach's value is estimated at 0.88, even when including students' perceptions on the learning of mathematics. This demonstrates a high level of reliability for the entire questionnaire, independent of the standards.

Standards	No. of items	Cronbach alpha	Reliability Level
Mathematics anxiety	5	.744	Acceptable
Confidence with mathematics	6	.663	Questionable
Students' perceptions on the learning of mathematics	2	.143	Unacceptable
The usefulness of mathematics	7	.804	Good
Enjoyment of mathematics.	3	.771	Acceptable
Social factors	7	.621	Questionable
Overall Cronbach's value	29	.881	Good

Table 4.2 Reliability of the questionnaire

4.3 Result of Mathematics Anxiety

<table>
<tr><td colspan="11">Mathematics Anxiety</td></tr>
<tr><th>Statement</th><th>Gender</th><th></th><th>Strongly disagree</th><th>Disagree</th><th>Unsure</th><th>Agree</th><th>Strongly agree</th><th>Mean</th><th>SD</th><th>Level</th></tr>
<tr><td rowspan="4">I worry about how well I am doing in mathematics.</td><td rowspan="2">Female</td><td>N</td><td>25</td><td>34</td><td>15</td><td>9</td><td>7</td><td rowspan="2">3.68</td><td rowspan="2">1.2</td><td rowspan="2">Disagree</td></tr>
<tr><td>%</td><td>27.8%</td><td>37.8%</td><td>16.7%</td><td>10%</td><td>7.8%</td></tr>
<tr><td rowspan="2">male</td><td>N</td><td>20</td><td>20</td><td>23</td><td>9</td><td>18</td><td>3.17</td><td>1.41</td><td>Unsure</td></tr>
<tr><td>%</td><td>22.2%</td><td>22.2%</td><td>25.6%</td><td>10%</td><td>20%</td><td></td><td></td><td></td></tr>
<tr><td rowspan="4">Taking mathematics tests scares me.</td><td rowspan="2">Female</td><td>N</td><td>32</td><td>32</td><td>12</td><td>7</td><td>7</td><td rowspan="2">3.83</td><td rowspan="2">1.22</td><td rowspan="2">Disagree</td></tr>
<tr><td>%</td><td>35.6%</td><td>35.6%</td><td>13.3%</td><td>7.8%</td><td>7.8%</td></tr>
<tr><td rowspan="2">male</td><td>N</td><td>21</td><td>16</td><td>25</td><td>16</td><td>12</td><td>3.20</td><td>1.34</td><td>Unsure</td></tr>
<tr><td>%</td><td>23.3%</td><td>17.8%</td><td>27.8%</td><td>17.8%</td><td>13.3%</td><td></td><td></td><td></td></tr>
<tr><td rowspan="4">I worry that other students might understand the problem better than me, when the teacher is showing the class how to do a problem.</td><td rowspan="2">Female</td><td>N</td><td>36</td><td>25</td><td>13</td><td>9</td><td>7</td><td rowspan="2">3.82</td><td rowspan="2">1.27</td><td rowspan="2">Disagree</td></tr>
<tr><td>%</td><td>40%</td><td>27.8%</td><td>14.4%</td><td>10%</td><td>7.8%</td></tr>
<tr><td rowspan="2">male</td><td>N</td><td>22</td><td>28</td><td>15</td><td>11</td><td>14</td><td>3.37</td><td>1.38</td><td>Disagree</td></tr>
<tr><td>%</td><td>24.4%</td><td>31.1%</td><td>16.7%</td><td>12.2%</td><td>15.6%</td><td></td><td></td><td></td></tr>
<tr><td rowspan="4">I worry about how well I am doing in mathematics more than other subjects.</td><td rowspan="2">Female</td><td>N</td><td>25</td><td>40</td><td>15</td><td>8</td><td>2</td><td rowspan="2">3.87</td><td rowspan="2">.997</td><td rowspan="2">Disagree</td></tr>
<tr><td>%</td><td>27.8%</td><td>44.4%</td><td>16.7%</td><td>8.9%</td><td>2.2%</td></tr>
<tr><td rowspan="2">male</td><td>N</td><td>24</td><td>16</td><td>25</td><td>15</td><td>10</td><td>3.32</td><td>1.33</td><td>Unsure</td></tr>
<tr><td>%</td><td>26.7%</td><td>17.8%</td><td>27.8%</td><td>16.7%</td><td>11.1%</td><td></td><td></td><td></td></tr>
<tr><td rowspan="4">When the teacher says he/she is going to ask me some questions to find out how much I know about mathematics, I worry that I will do poorly.</td><td rowspan="2">Female</td><td>N</td><td>18</td><td>21</td><td>14</td><td>21</td><td>16</td><td rowspan="2">3.04</td><td rowspan="2">1.41</td><td rowspan="2">Unsure</td></tr>
<tr><td>%</td><td>20%</td><td>23%</td><td>15.6%</td><td>23.3%</td><td>17.8%</td></tr>
<tr><td rowspan="2">male</td><td>N</td><td>13</td><td>14</td><td>19</td><td>25</td><td>19</td><td>2.74</td><td>1.34</td><td>Unsure</td></tr>
<tr><td>%</td><td>14.4%</td><td>15.6%</td><td>21.1%</td><td>27.8%</td><td>21.1%</td><td></td><td></td><td></td></tr>
<tr><td></td><td>Female</td><td colspan="5">Weighted mean = 2.3511</td><td colspan="3">Std.Deviation= .79000</td></tr>
<tr><td></td><td>male</td><td colspan="5">Weighted mean = 2.8400</td><td colspan="3">Std.Deviation= .99310</td></tr>
</table>

Table 4.3: Summary of participants response on mathematic anxiety

Table 4.3 shows the mathematics anxiety of students. Reverse coding of negative expressions was not used for separate analysis of the statements separately, so high scores represent low anxiety, while a low score means high anxiety (See table 4.1). Females and males did not show any high anxiety level for all statements in table 4.3.

Male students showed a moderate anxiety, on the following statements:

- "I worry about how well I am doing in mathematics (M = 3.17, SD = 1.41).

- " Taking mathematics tests scares me" (M = 3.20, SD =1.34").
- "I worry about how well I am doing in mathematics more than other subjects" (M = 3.32, SD = 1.33).

In contrast, it can be seen from table 4.3 that female students showed low anxiety levels, on the following statements:

- "I worry about how well I am doing in mathematics" (M = 3.68, SD =1.2)
- " Taking mathematics tests scares me" (M = 3.83, SD =1.22).
- "I worry about how well I am doing in mathematics more than other subjects" (M = 3.87, SD =.997).

Both females and males scored moderate on the anxiety level, when responding to the statement "When the teacher says he/she is going to ask me some questions to find out how much I know about mathematics, I worry that I will do poorly." (Females, M = 3.04, SD =1.41), (Males, M = 2.74, SD = 1.34), while they both scored low on the anxiety level, (Males, M = 3.37, SD =1.38) (Females, M = 3.82, SD =1.27), when responding to the statement "I worry that other students might understand the problem better than me, when the teacher is showing the class how to do a problem."

Total anxiety level score, after reverse coding for all negative expressions, it can be seen from demonstrate lower anxiety levels for females, (Weighted mean = 2.35, SD = .79), while males students showed moderate anxiety, (Weighted mean = 2.84, SD =.993). T-test was applied to compare the means between males and females, revealing significant differences between male and female (t=-3.655, df= 169.4, p < 0.001).

Two main points emerge from the above results. Firstly, in all statements, females showed low anxiety levels, except when asked questions to ascertain how much they knew about mathematics. Here both females and males just showed moderate anxiety, because of concerns about performing poorly. This worry, which is cognitive and concerns performance and the effects of failing (Wigfield and Meece, 1988), may affect their attitude towards mathematics (Tahar et al., 2010; Dowker et al., 2016, Belbase, 2010). Secondly, although the weight of research evidence suggests that girls have greater mathematics anxiety (Dowker et al, 2016), Table 4.3 shows that female students showed lower anxiety levels compared with males. This is not easy to interpret but perhaps reflects different emphases in cultural environment and educational context, due to gender-based separation at all levels (Sait et al.,2004). This concurs with Norton and Rennie (1998)'s findings in 13-16-year olds in Australia, that girls from coeducational schools report more anxiety than girls in the single-sex schools. However, it is the view of the current researcher that girls have greater mathematics anxiety in a co-educational situation, where reluctance to answer questions may be due to fear of mockery by the boys. In an all female class with a female teacher they may feel more relaxed about responding: a potential argument in favour of single sex schools.

4.4 Result of Confidence with Mathematics

Confidence with mathematics									
Statement	Gender		Strongly disagree	Disagree	Unsure	Agree	Strongly agree	Mean	SD	Level
Mathematics is more difficult for me than for many of my classmates.	Female	N	56	26	6	2	0	4.51	.723	Strongly disagree
		%	62.2%	28.9%	6.7%	2.2%	0%			
	male	N	35	25	14	10	6	3.81	1.25	Disagree
		%	38.9%	27.8%	15.6%	11.1%	6.7%			
I'm confident I can understand the basic concepts taught in my mathematics class	Female	N	1	0	7	26	56	1.49	.738	Strongly agree
		%	1.1%	0%	7.8%	28.9%	62.2%			
	male	N	4	1	10	29	46	1.76	1.009	Strongly agree
		%	4.4%	1.1%	11.1%	32.2%	51.1%			
I am good at using mathematics to solve real-life problems	Female	N	7	11	23	36	13	2.59	1.12	Agree
		%	7.8%	12.2%	25.6%	40%	14.4%			
	male	N	9	11	29	30	11	2.74	1.13	Unsure
		%	10%	12.2%	32.2%	33.3%	12.2%			
I find mathematics easy.	Female	N	2	4	17	22	45	1.84	1.02	Agree
		%	2.2%	4.4%	18.9%	24.4%	50%			
	male	N	7	10	23	26	24	2.44	1.21	Agree
		%	7.8%	11.1%	25.6%	28.9%	26.7%			
My teacher thinks I can do well in mathematics.	Female	N	1	3	8	32	46	1.68	.859	Strongly agree
		%	1.1%	3.3%	8.9%	35.6%	51.1%			
	male	N	0	0	6	27	63.3	1.43	.619	Strongly agree
		%	0%	0%	6.7%	30%	63.3%			
I am confident in solving mathematics problems.	Female	N	0	3	20	36	31	1.94	.839	Agree
		%	0%	3.3%	22.2%	40%	34.4%			
	male	N	0	3	21	32	34	1.92	.864	Agree
		%	0%	3.3%	23.3%	35.6%	37.8%			
	Female	Weighted mean = 1.8389				Std.Deviation= .54306				
	male	Weighted mean = 2.0815				Std.Deviation= .63954				

Table 4.4: Summary of participants response on Confidence with Mathematics

Table 4.4 shows confidence with mathematics for students. All statements are positive expressions except " Mathematics is more difficult for me than for many of my classmates", I did not conduct reverse coding for this statement. Therefore, high scores on negative statement represent high confidence, while a low score means low confidence (See table 4.1).

Females and males did not show low confidence levels for any statements (table 4.4). Males showed moderate confidence level for the statement "I am good at using mathematics to solve real-life problems". (M = 2.74, SD = 1.13), and high confidence in the statements "Mathematics is more difficult for me than for many of my classmates" (M = 3.81, SD =1.25).

In contrast, females scored very highly for the statement "Mathematics is more difficult for me than for many of my classmates" (M = 4.51, SD = .723), and showed high confidence in the statement "I am good at using mathematics to solve real-life problems" (M = 2.59, SD = 1.12).

Both females and males show very high confidence in the statements " My teacher thinks I can do well in mathematics" (Males, M = 1.43, SD = .619), (Females, M = 1.68, SD = .859), " I'm confident I can understand the basic concepts taught in my mathematics class" (Males, M = 1.76, SD = 1.009), (Females, M = 1.49, SD = .738). Moreover, females and male students showed high confidence in the statements "I am confident in solving mathematics problems" (Males, M = 1.92, SD = .864), (Females, M = 1.92, SD = .864), " I find mathematics easy" (Males, M = 2.44, SD = 1.21), (Females, M = 1.84, SD = 1.02).

For total confidence with mathematics level score, after reverse coding for negative expression, it can be seen from table 4.4 that both females and males show high confidence toward mathematics. However, females show slightly higher confidence (Weighted mean = 1.83, SD = .543) toward mathematics than males (Weighted mean = 2.08, SD = .639). T-test was applied to compare the means between males and female students. The tests revealed that there is significant difference between male and female (t=-2.743, df= 174.44, p = 0.007).

Finally, based on the results above, two main points emerge. Firstly, although both females and males showed high levels of confidence about mathematics, out of 39 countries, Saudi middle schools ranked lowest in their TIMSS mathematics performance in 2015 an ongoing downward trend since 2011. According to result of TIMMS in 2015, the majority of Saudi students (62%) were very confident or confident in mathematics, but 38% were not confident. Therefore, there is gap between performance and confidence about mathematics, with high confidence but low performance, which I expect to be borne out if participants in my study take part in TIMMS in 2019.

However, (according to my experience as a teacher) I do not expect an achievement/confidence gap for class exams because some of Saudi middle school teachers promote procedural knowledge (Skemp, 1976; Star and Stylianides, 2013) in the following manner:

1-   Teacher gives students the rule, and students copy this rule.

2-   Teacher explains the rule using an example and students copy this example.

3-   Students work other examples which are similar to first example.

4-   Teacher gives students next rule and so on.

Then, in the final exam, I have noticed that teachers give students questions which are the same examples that students were given in class. For example, if the example in the class was, "find the area and perimeter of shape below".

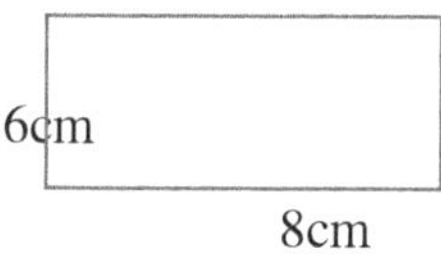

The question in exam will be like this example. Sometimes even the numbers will be the same. Thus, students might obtain high grades in mathematics which in turn increases their confidence which is vital to mathematics success (Nurmi et al., 2003).

However, I am concerned that, when students face more conceptual-based tests such as TIMMS, they will perform poorly and risk losing their confidence in mathematics, which in turn could affect their attitude towards mathematics. Therefore, a good teacher should combine procedural and conceptual knowledge.

Secondly, Table 4.4 shows that females tend to be more confident in mathematics than males. This raises two points:

Firstly, as mentioned previously, this might be attributable to the cultural educational context. For example, US study of middle school students' attitudes and behaviour found a positive impact for the single-sex setting for girls' mathematics learning, their asking of mathematics questions and how they see themselves as mathematicians (Streitmatter, 1997).

Secondly, in my study, high confidence in mathematics in girls could be associated with their low anxiety levels, which is consistent with prior research findings across recent decades (Parsons et al., 2009) (see chapter 2).

4.5. Result of Perceived Usefulness of Mathematics

Perceived Usefulness of Mathematics										
Statement	**Gender**		**Strongly disagree**	**Disagree**	**Unsure**	**Agree**	**Strongly agree**	**Mean**	**SD**	**Level**
Mathematics teaches me to think clearly.	Female	N	0	3	5	44	38	1.70	.726	Strongly agree
		%	0%	3.3%	5.6%	48.9%	42.2%			
	male	N	2	1	14	47	26	1.96	.833	Agree
		%	2.2%	1.1%	11.1%	52.2%	28.9%			
Mathematics is related in my life.	Female	N	8	8	15	23	36	2.21	1.30	Agree
		%	8.9%	8.9%	16.7%	25.6%	40%			
	male	N	10	15	17	25	23	2.60	1.33	Unsure
		%	11.1%	16.7%	18.9%	27.8%	25.6%			
Mathematics is useful for my career.	Female	N	3	3	4	25	55	1.60	.969	Strongly agree
		%	3.3%	3.3%	4.4%	27.8%	61.1%			
	male	N	6	3	9	18	54	1.77	1.18	Strongly agree
		%	6.7%	3.3%	10%	20%	60%			
I think mathematics is useful only for tests.	Female	N	55	21	7	5	2	4.36	.998	Strongly disagree
		%	61.1%	23.3%	7.8%	5.6%	2.2%			
	male	N	51	22	10	4	3	4.27	1.04	Strongly disagree
		%	56.7%	24.4%	11.1%	4.4%	3.3%			
Mathematics helps me to understand reports and advertisements about prices, sale, percentages etc.	Female	N	3	0	4	20	63	1.44	.863	Strongly agree
		%	3.3%	0%	4.4%	22.2%	70%			
	male	N	5	2	3	17	63	1.54	1.06	Strongly agree
		%	5.6%	2.2%	3.3%	18.9%	70%			
Mathematics is an important subject for me.	Female	N	1	2	9	25	53	1.59	.847	Strongly agree
		%	1.1%	2.2%	10%	27.8%	58.9%			
	male	N	5	4	14	23	44	1.92	1.15	Agree
		%	5.6%	4.4%	15.6%	25.6%	48.9%			
I think mathematics is useful in solving real world problems.	Female	N	4	12	33	25	16	2.59	1.06	Agree
		%	4.4%	13.3%	36.7%	27.8%	17.8%			
	male	N	5	16	28	21	20	2.61	1.17	Unsure
		%	5.6%	17.8%	31.1%	23.3%	22.2%			
	Female	Weighted mean = 1.8254				Std.Deviation= .66621				
	male	Weighted mean = 2.0190				Std.Deviation= .75538				

Table 4.5: Summary of participants' responses on Perceived Usefulness of Mathematics.

Table 4.5 shows the perceived usefulness of mathematics. To analyse the statements separately, all statements are positive expressions except, " I think mathematics is useful only for tests.". I did not conduct reverse coding for this question. Therefore, high scores on this question represent positive perceptions about usefulness of Mathematics, while a low score means negative perceptions about usefulness of Mathematics (See table 4.1). There were no negative perceptions among female and male students about usefulness of mathematics in the real world for all statements (see table 4.5).

Females showed high positive perceptions about the usefulness of mathematics for the statement, "Mathematics teaches me to think clearly." (M = 1.70, SD = .726), "Mathematics is an important subject for me" (M = 1.59, SD =.847). Furthermore, females scored positive perceptions in two statements which are:

- "Mathematics is related in my life." (M = 2.21, SD = 1.30).
- " I think mathematics is useful in solving real world problems." (M = 2.59, SD =1.06).

In contrast, males showed moderate perceptions about usefulness of Mathematics, when they responded to the statement of "Mathematics is related in my life."(M = 2.60, SD =1.33), "I think mathematics is useful in solving real world problems" (M = 2.61, SD =1.17). Furthermore, they scored positive perceptions in two statements, "Mathematics teaches me to think clearly." (M = 1.96, SD =.833), "Mathematics is an important subject for me" (M = 1.92, SD =1.15).

Both males and females showed high positive perceptions about usefulness of Mathematics for statements:
- "Mathematics is useful for my career" (Females, M = 1.60, SD =.969), (Males, M = 1.77, SD =1.18).
- "I think mathematics is useful only for tests" (Females, M = 4.36, SD =.998), (Males, M = 4.27, SD =1.04).
- "Mathematics helps me to understand reports and advertisements about prices, sale, percentages etc" (Females, M = 1.44, SD =.863), (Males, M = 1.54, SD =1.06).

For total usefulness of mathematics level score, after reverse coding for all negative expressions, it can be seen that boys and girls hold positive views regarding usefulness of mathematics. Although, females show slightly high positive perceptions about usefulness of Mathematics (Weighted mean = 1.8254, SD = .66621) compared with males, (Weighted mean =2.0190, SD =.75538), the results indicate no significant differences between females and males (t=-1.824, df= 175.26, p = 0.07).

Finally, based on results above, despite the suggestions by other studies that girls hold negative perceptions (Sherman,1978), in this study both genders recognise the relevance of mathematics to their daily lives and futures, and there is no statistically significant gender differences. This belief is in line with the Adelson and McCoach (2011, p.226)'s definition of the perceived usefulness of mathematics (see chapter 2). Therefore, for the students in the current study, positive perceptions may increase motivation to study and practice mathematics (Pajares & Miller, 1994). (See section 4.9 in this chapter).

4.6. Result of Enjoyment of Mathematics.

Enjoyment of Mathematics									
Statement	Gender		Strongly disagree	Disagree	Unsure	Agree	Strongly agree	Mean	SD	Level
I am enjoying doing mathematics.	Female	N	1	5	13	22	49	1.74	.978	Strongly agree
		%	1.1%	5.6%	14.4%	24.4%	54.4%			
	male	N	7	4	27	28	24	2.36	1.15	Agree
		%	7.8%	4.4%	30%	31.1%	26.7%			
I find mathematics interesting and motivating.	Female	N	3	6	10	29	42	1.88	1.06	Agree
		%	3.3%	6.7%	11.1%	32.2%	46.7%			
	male	N	6	5	24	33	22	2.33	1.11	Agree
		%	6.7%	5.6%	26.7%	36.7%	24.4%			
I find mathematics boring.	Female	N	42	26	14	3	5	4.08	1.12	Disagree
		%	46.7%	28.9%	15.6%	3.3%	5.6%			
	male	N	27	25	18	10	10	3.54	1.32	Disagree
		%	30%	27.8%	20%	11.1%	11.1%			
	Female	Weighted mean = 1.8481				Std.Deviation= .87718				
	male	Weighted mean = 2.3815				Std.Deviation= .97222				

Table 4.6: Summary of participants' responses on enjoyment of mathematics.

Table 4.6 shows result of enjoyment of mathematics. All statements are positive expressions except "I find mathematics boring." I did not conduct reverse coding for this question. Therefore, high scores on this question represent high enjoyment in mathematics, while a low score means low enjoyment of mathematics (See table 4.1). Females showed very high scores for the statement " I am enjoying doing mathematics." (M = 1.74, SD = .978), while males have high scores in this statement (M = 2.36, SD = 1.15).

Both males and females had high scores for the statements:

- "I find mathematics interesting and motivating." (Females, M = 1.88, SD =1.06), (Males, M = 2.33, SD =1.11).
- "I think mathematics is useful only for tests" (Females, M = 4.08, SD =1.12), (Males, M = 3.54, SD =1.32).

For total enjoyment of mathematics level score, after reverse coding "I find mathematics boring.", it can be seen from the table that females have slightly more enjoyment of mathematics than males. (Females, Weighted mean = 1.8481, SD =.87718), (Males, Weighted mean= 2.3815, SD=.97222). The results show significant differences between female and male students in terms of enjoyment of mathematics (t=-3.86, df= 176.14, p = 0.007).

Two main points emerge from these results. Firstly, as depicted in Table 4.6, the weighted mean was below 2.60, suggesting that female and male students really enjoyed mathematics, an important indicator of intrinsic motivation (Ryan and Deci, 2000), encouraging study for its own sake. Secondly, females experience slightly more enjoyment of mathematics than males, in line with Leder and Forgasz (2002)'s findings that, in Australia, female middle school students had higher levels of interest in and enjoyment of mathematics. It contrasts with PISA (2012)'s findings, published by OECD (2013), that male students were significantly more interested in mathematics and enjoyed it more than girls. This could relate to the cultural environmental and educational context. For example, an Irish study by Prendergast and O'Donoghue (2014) found that the two top ranking schools in terms of students' enjoyment of mathematics were both single-sex schools. The latter findings support Saudi single-sex educational policy.

4.7. Results of mathematics class environment relationship with teacher and peers.

<table>
<tr><td colspan="11" align="center">Mathematics class environment relationship with teacher and peers</td></tr>
<tr><td>Statement</td><td>Gender</td><td></td><td>Strongly disagree</td><td>Disagree</td><td>Unsure</td><td>Agree</td><td>Strongly agree</td><td>Mean</td><td>SD</td><td>Level</td></tr>
<tr><td rowspan="4">My mathematics teacher encourages me at times when I don't do well in class.</td><td rowspan="2">Female</td><td>N</td><td>6</td><td>3</td><td>13</td><td>34</td><td>34</td><td rowspan="2">2.03</td><td rowspan="2">1.12</td><td rowspan="2">Agree</td></tr>
<tr><td>%</td><td>6.7%</td><td>3.3%</td><td>14.4%</td><td>37.8%</td><td>37.8%</td></tr>
<tr><td rowspan="2">male</td><td>N</td><td>5</td><td>5</td><td>13</td><td>25</td><td>42</td><td>1.96</td><td>1.16</td><td>Agree</td></tr>
<tr><td>%</td><td>5.6%</td><td>5.6%</td><td>14.4%</td><td>27.8%</td><td>46.7%</td></tr>
<tr><td rowspan="4">Other students in my mathematics class take my ideas seriously.</td><td rowspan="2">Female</td><td>N</td><td>7</td><td>6</td><td>30</td><td>30</td><td>17</td><td rowspan="2">2.51</td><td rowspan="2">1.11</td><td rowspan="2">Agree</td></tr>
<tr><td>%</td><td>7.8%</td><td>6.7%</td><td>33.3%</td><td>33.3%</td><td>18.9%</td></tr>
<tr><td rowspan="2">male</td><td>N</td><td>13</td><td>18</td><td>29</td><td>19</td><td>11</td><td>3.03</td><td>1.22</td><td>Unsure</td></tr>
<tr><td>%</td><td>14.4%</td><td>20%</td><td>32.2%</td><td>21.1%</td><td>12.2%</td></tr>
<tr><td rowspan="4">Even thinking about my mathematics class makes me feel hopeless.</td><td rowspan="2">Female</td><td>N</td><td>53</td><td>24</td><td>8</td><td>3</td><td>2</td><td rowspan="2">4.37</td><td rowspan="2">.942</td><td rowspan="2">Strongly disagree</td></tr>
<tr><td>%</td><td>58.9%</td><td>26.7%</td><td>8.9%</td><td>3.3%</td><td>2.2%</td></tr>
<tr><td rowspan="2">male</td><td>N</td><td>33</td><td>28</td><td>17</td><td>7</td><td>5</td><td>3.86</td><td>1.16</td><td>Disagree</td></tr>
<tr><td>%</td><td>36.7%</td><td>31.1%</td><td>18.9%</td><td>7.8%</td><td>5.6%</td></tr>
<tr><td rowspan="4">I enjoy participating in this mathematics class.</td><td rowspan="2">Female</td><td>N</td><td>1</td><td>2</td><td>3</td><td>26</td><td>58</td><td rowspan="2">1.47</td><td rowspan="2">.767</td><td rowspan="2">Strongly agree</td></tr>
<tr><td>%</td><td>1.1%</td><td>2.2%</td><td>3.3%</td><td>28.9%</td><td>64.4%</td></tr>
<tr><td rowspan="2">male</td><td>N</td><td>2</td><td>5</td><td>16</td><td>25</td><td>42</td><td>1.89</td><td>1.03</td><td>Agree</td></tr>
<tr><td>%</td><td>2.2%</td><td>5.6%</td><td>17.8%</td><td>27.8%</td><td>46.7%</td></tr>
<tr><td rowspan="4">My mathematics teacher respects me.</td><td rowspan="2">Female</td><td>N</td><td>2</td><td>3</td><td>6</td><td>20</td><td>59</td><td rowspan="2">1.54</td><td rowspan="2">.926</td><td rowspan="2">Strongly agree</td></tr>
<tr><td>%</td><td>2.2%</td><td>3.3%</td><td>6.7%</td><td>22.2%</td><td>65.6%</td></tr>
<tr><td rowspan="2">male</td><td>N</td><td>0</td><td>0</td><td>3</td><td>14</td><td>73</td><td>1.22</td><td>.492</td><td>Strongly agree</td></tr>
<tr><td>%</td><td>0%</td><td>0%</td><td>3.3%</td><td>15.6%</td><td>81.1%</td></tr>
<tr><td rowspan="4">I feel like my presence matters in my mathematics class.</td><td rowspan="2">Female</td><td>N</td><td>0</td><td>3</td><td>18</td><td>32</td><td>37</td><td rowspan="2">1.86</td><td rowspan="2">.855</td><td rowspan="2">Agree</td></tr>
<tr><td>%</td><td>0%</td><td>3.3%</td><td>20%</td><td>35.6%</td><td>41.1%</td></tr>
<tr><td rowspan="2">male</td><td>N</td><td>5</td><td>10</td><td>30</td><td>23</td><td>22</td><td>2.48</td><td>1.14</td><td>Agree</td></tr>
<tr><td>%</td><td>5.6%</td><td>11.1%</td><td>33.3%</td><td>25.6%</td><td>24.4%</td></tr>
<tr><td></td><td>Female</td><td colspan="5">Weighted mean = 1.8407</td><td colspan="3">Std.Deviation= .57459</td></tr>
<tr><td></td><td>male</td><td colspan="5">Weighted mean = 2.1204</td><td colspan="3">Std.Deviation= .58006</td></tr>
</table>

Table 4.7: Summary of participants' responses on mathematics class environment relationship with teacher and peers.

Table 4.7 shows the satisfaction of students in the Mathematics class environment, which is related to teachers and peers. All statements are positive expressions except "Even thinking about my mathematics class makes me feel hopeless". I did not conduct reverse coding for this question (See table 4.1). All students expressed satisfaction for all statements in this section. (table 4.7).

Females showed more satisfaction for two statements " I enjoy participating in this mathematics class." (M = 1.47, SD = .767), and "Even thinking about my mathematics class makes me feel hopeless"(M = 4.37, SD = .942). While they showed satisfaction for the statement, "Other students in my mathematics class take my ideas seriously." (M = 2.51, SD = 1.11).

In contrast, males showed satisfaction for two the statements, "Even thinking about my mathematics class makes me feel hopeless" (M = 3.86, SD = 1.16), and, "I enjoy participating in this mathematics class" (M = 1.89, SD = 1.03). While they showed moderate levels of satisfaction for statement, " Other students in my mathematics class take my ideas seriously" (M = 3.03, SD = 1.22).

Both females and males showed their satisfaction, when they responded to the statements:

- "My mathematics teacher encourages me at times when I don't do well in class." (Females, M = 2.03, SD =1.12), (Males, M = 1.96, SD =1.16).
- "I feel like my presence matters in my mathematics class." (Females, M = 1.86, SD =.855), (Males, M = 2.48, SD =1.14).

They both showed more satisfaction for statement, " My mathematics teacher respects me" (Females, M = 1.54, SD =.926), (Males, M = 1.22, SD =.492).

When examining the total for all statements, I reversed coding for the negative expression. The table shows that females were more satisfied than boys with the mathematics class environment, teachers and peers (Females, Weighted mean = 1.8407, SD = .57459), (Males, Weighted mean = 2.1204, SD =.58006). T-test was applied to compare the means between males and female students. The test revealed a significant difference between males and females (t=-3.24, df= 177.984, p = 0.001).

Finally, based on results above, two main points can be made. Firstly, both females and males showed satisfaction with their teacher, particularly for the statement, "My mathematics teacher respects me". Their positive relationship with their teacher will affect the mathematics classrooms and hence their attitude toward mathematics (Sakiz et al., 2012; Newman and Schwager, 1993; Orton and Wain, 1994). Moreover, all students experience a sense of belonging to the mathematics classrooms, because they relate positively with their peers (Goodenow, 1993a), which will also affect their attitude toward mathematics, especially, as peers play an important role in development of mathematics attitudes during adolescence (e.g., Berndt 1979). Secondly, females were more satisfied than boys with the mathematics class environment, related to teacher. This might be attributable to the cultural environment and educational context. For example, in Nigeria, Lee and Lockheed (1990), found that girls had more positive attitudes to mathematics if taught by women teachers. However, in middle school, the gender of teachers had no impact on students' grading behaviours in some studies (e.g. Wiles,1992).

4.8 Overall result of students' attitudes towards mathematics.

Attitude towards mathematics	*Female*	Weighted mean = 2.0479	Std.Deviation= .48551
Attitude towards mathematics	*male*	Weighted mean = 2.3368	Std.Deviation= .54737

Table 4.8: Summary of participants' responses on attitude toward mathematics.

Table 4.8. shows overall result of students' attitudes towards mathematics in the following areas:

1- Mathematics anxiety.

2- Confidence with mathematics.

3- Students' perceptions on the learning of mathematics.

4- The usefulness of mathematics.

5- Enjoyment of mathematics.

6- Mathematics class environment and relationship with teacher and peers.

After reverse coding for all negative expressions, it can be seen from table 4.8 that more females than males show positive attitudes toward mathematics. (Females, Weighted mean = 2.0479, SD = .48551), (Males, Weighted mean = 2.3368, SD =.54737). However, both male and female students have almost the same attitude toward mathematics. T-test was applied to compare the means between males and female students. The test revealed that there is significant difference between male and female (t=-3.746, df= 175.501, p < 0.001).

Based on results above, although, studies suggest that boys' attitudes are generally more positive than girls' (Frost et al., 1994; Foxman et al., 1981, Else-Questet al, 2013), table 4.8 shows that female students had more positive attitudes toward mathematics than their male counterparts. This is not easy to interpret. The explanation for the findings of the current study can be attributed three reasons. Firstly, they can reflect different emphases in the cultural environment and educational context, from country to country (e.g. Cogan and Schmidt, 1999). For example, Reilly et al (2019), reporting results from the 2011 (TIMSS), indicate that boys showed more positive attitudes than girls in most countries, with the exception of some Middle Eastern countries (e.g., Oman) where this trend was reversed. Secondly, the explanation for the findings of the current study can be attributed the gender-role socialization pattern in Saudi culture in line with the findings of Birenbaum and Nasser (2006) about the influences of gender and ethnicity on attitudes towards mathematics for Jewish and Arab eighth graders in Israel (see chapter 2, section 3). Thirdly, one other explanation could be the strict gender segregation in Saudi educational policy. For example, in his study of attitude relating to gender separated schools in Australia in secondary schools (including age 13 to 16), Norton and Rennie (1998), generally, found that girls from single-sex schools report more positive attitudes toward mathematics than girls in the coeducational schools. However, Reilly et al (2019) found mixed support for the gender segregation hypothesis.

Finally, both male and female students have positive attitudes toward mathematics. Beaton et al (1996) found that it is often possible to observe a positive connection between achievement in mathematics and students' attitudes towards the subject. Therefore, it might be that the students in the current study had high achievement in mathematics in class. However, I am expecting that they probably will have lower achievement in TIMMS, if participants in my study take part in next TIMMS in 2019 (see section 4.4 in this chapter). If this is the case, they will have positive attitudes towards mathematics, but have lower achievement. Papanastasiou (2000) attributed this anomaly to teachers' low expectations, which are easy for students to satisfy and leads to the development of a positive attitude towards mathematics. In other words, Saudi teachers may want from their students get high achievement in mathematics in the exam class, through giving them easy exams in mathematics (see section 4.4 in this chapter), which in turn causes them to develop positive attitudes towards mathematics.

However, Saudi teachers may not be fostering a good future in mathematics for their students, as evidenced by the fact that Saudi Arabia has continued to take part in TIMMS, getting low rankings in mathematics for middle schools.

4.9 Analysis the open-ended questions.

A qualitative method, with thematic analysis, was employed to analyse open-ended questions.

Question one: an open-ended question asked girl and boy students to report on their perceptions of why they liked, or disliked mathematics. 87 out of 90 of the girls' responses to this question were analysed and 85 out of 90 of the boys' responses. 3 responses from females and 5 responses from males were excluded because they were unclear. 79 out of 87 of females like mathematics, while 72 out of 85 of males like mathematics. In contrast, 8 out of 87 of females dislike mathematics, while 13 out of 85 of males dislike mathematics. Coding of this question resulted in five interpretable themes for both females and males, as shown, in order of importance in the table 4.9. There were themes which appeared only for girls, which are teachers on the internet and social media personalities, links between mathematics and the Islamic religion, and Mathematicians. One theme appeared only for boys, which is the encouragement and support received from parents. (I differentiated thematically between some answers, where multiple responses were given. For example, some students said I like mathematics because it is useful for my daily life and my career, but also said I like it because of my teacher and it is important in my life)

Factors affecting the females' attitudes about learning of mathematics	Factors affecting the males' attitudes about learning of mathematics
1- Usefulness of mathematics in everyday life.	1- Usefulness of mathematics in everyday life.
2- The importance of the teacher (positive and negative)	2- The importance of the teacher (positive).
3- Teachers on the internet and social media personalities (positive).	3- The notable difficulty of mathematics.
4- Mathematics for career and 'future life'	4- Enjoyment of mathematics.
5- Enjoyment of mathematics.	5- Mathematics for career and 'future life'.
6- The links between mathematics and the Islamic religion, and mathematicians (positive).	6- The encouragement and support received from parents.
7- The notable difficulty of mathematics.	

Table 4.9: Summary of students' perception of why they liked, or disliked mathematics,

Usefulness of mathematics in everyday life:

The most common reasons for perceptions of why students liked or disliked mathematics was the usefulness of mathematics in everyday life. 56 out of 87 girls and 33 out of 85 boys, liked mathematics because of its usefulness. They said that mathematics helps us for most of the things in everyday life. One male respondent illustrated this finding by saying, "Mathematics helps me to understand prices, sales, percentages", while one of the girls said, " Mathematics is useful for my daily life, for example, when I want to calculate the area of a room."

It might be that this theme is the most common because of Saudi Arabia's new curriculum, (detailed in chapter one) which makes intensive use of examples related to daily life (McGraw-Hill, 2019), linking mathematics to reality in line with the theory of Realistic Mathematics Education (RME). The philosophy underpinning RME is "that students should develop their mathematical understanding by working from contexts that make sense to them" (Dickinson and Hough, 2012:1). According to Freudenthal's (1977) theory of RME, mathematics must be connected to reality. Therefore, when a student recognises that mathematics is related to their life, this idea could improve their attitude towards mathematics. This argument is supported by a number of studies which refer to the theory that Realistic Mathematics Education positively impacts students' attitude levels towards mathematics (Devrim and Uyangor, 2006).

The importance of the teacher:

The second most common reasons for perceptions of why girls and boys liked, or disliked mathematics was the importance of the teacher. 35 out of 87 of the female students liked mathematics, and 3 out of 87 of the female students disliked mathematics because of its teacher. In contrast, 29 of 85 of the male students liked mathematics because of their teacher.

The importance of the teacher in helping students to like mathematics was stressed by the female and male respondents. It is essential for students to have a teacher who makes lessons enjoyable, ensures that students understand the work and is a kind and respectful person. These factors help foster a positive attitude towards mathematics. This is evidenced by quotes from three respondents. The first, by a boy was, " I like it because of my teacher, and I believe that the main reason to like mathematics is the teacher". The second, from one of the girls, " I like mathematics because my teacher a kind person, she loves us and we love her; she respects us and we respect her." The third, from a boy, " I like it because my teacher respects me". There are many studies which have indicated that student attitudes toward study, teachers, methods, and the overall school climate are influenced by the teacher-student relationship (e.g. Torrance et al., 1966 ; Fraser and Fisher, 1982; Hartmut, 1978). Therefore, teachers may be blamed for making students hate mathematics (Aiken and Dreger, 1961).This is evidenced by the comments of one girl respondent, who said, " I liked mathematics in the past, but now I hate mathematics because of my teacher, her explanations and her teaching style".

Finally, as mentioned in chapter 2, the work of Roeser et al. (1998) supports the view that the teachers greatly affect the achievement and emotions of students in middle school, with lasting consequences. This is evidenced by the quotes above, which underline the key role played by respect from teachers in students' attitude toward mathematics.

Mathematics for career and 'future life':

The fourth most common reasons for girls and the fifth most common reasons for boys liking of mathematics was career prospects and future life. 15 out of 87 of the female students liked mathematics because of its importance in their future. In contrast, 9 out of 85 of the male students liked mathematics because of its importance in their future. They expected that they would study mathematics at university, or they dreamed of a career requiring mathematics. One female respondent illustrated this finding by saying, "I like mathematics because it will help me realize my dream of becoming an astronomer", while a boy said, "I like mathematics because I want to study Computer Science in the future life, so I need mathematics".

Dr. Judy Willis (2010), who has combined the knowledge and experience gained through her dual careers as a maths teacher and a neurologist, suggests that a love of maths can be fostered through its connection to students' personal interests and targets. This is illustrated in the current study, since respondents who like mathematics, related it to their future goals.

Enjoyment of mathematics:

Enjoyment of mathematics: the fifth most common reasons for girls and the fourth most common reasons for boys was career prospects and future life. 9 out of 87 of the female students, and 10 out of 85 of the male students, liked mathematics because they enjoyed it. One female respondent illustrated this finding by saying, " I liked mathematics because it is interesting and enjoyable". A male student said, " I like mathematics because it is enjoyable and has ambiguity that makes you enthusiastic about solving its problems ". (see also "4.6. Result of Enjoyment of Mathematics").

The Notable Difficulty of mathematics:

The most common reasons for perceptions of why girl and boys disliked mathematics was the mathematics is notably difficult. 5 out of 87 of the female students, and 13 out of 85 disliked mathematics because its difficulty. This is evidenced by quotes from one respondent. The first, by a girl was, "I dislike mathematics. Mathematics is difficult because it depends on understanding, and I prefer memorising". The second, from one of the boys, " I dislike mathematics because it is complicated and difficult".

Unfortunately, mathematics is regarded by many students as a difficult, complicated and abstract subject (Sharples, 1969; Dossey et.al, 1988). Such negative attitudes towards mathematics may be detrimental to the learning process. Although, girls have a greater tendency than boys to believe that mathematic is difficult (Foxman et al., 1982), in my study, boys tend to believe, more than do girls, that mathematics is difficult. This finding is in line with Leder and Forgasz (2002)'s findings, mentioned in chapter 2. The difficulties of mathematics generally, they could be due to many reasons such as teacher style, curriculum, ...etc.

Teachers on the internet and social media personalities:

One of the most striking results to emerge from this study was the impact of social media personalities and internet teachers on girls' attitudes to mathematics. It was the third most common reasons for perceptions of why girls liked mathematics. However, I had not expected this to be a factor at all in student perceptions. 18 out of 87 of the female students liked mathematics because of social media. This is evidenced by quotes from two respondents. One stated, "I like mathematics because one of the social media personalities on " Snapchat" was talking about the benefits of mathematics". The second said, "I am following one of the social media personalities, and from his style and his speaking about mathematics, I liked mathematics."

The greatest social media consumers are adolescents (Booker et al., 2018). In Saudi Arabia, most users of Facebook, Twitter and Instagram and Twitter were male. However, amongst women, Snapchat, the messaging app was more popular (Arab News, 2019). Figure 4.1 presents a ranking of the countries with the largest Snapchat audiences worldwide as of July 2019. During the measured period, Saudi Arabia were ranked forth with an audience of 15.4 million users (Statista, 2019).

These statistics indicated an effect of Snapchat personalities on today's generation in Saudi Arabia, especially girls. Moreover, I found that respondents in level 3 (age 15-16), might know one of the social media personalities on Snapchat, who speak positively about mathematics and that they spread their name among them. Regardless of the type of social media in question (whether it is Snapchat, Twitter or any other app), research now suggests that media has an important effect on students' image of mathematics which in turn affects their attitudes towards mathematics. A significant study in this field is the work done by Nardi (2016), who suggests that as a result of this effect, it is important for teachers to be aware of such media personalities and to learn to capitalise on the potential positive effect of media personalities

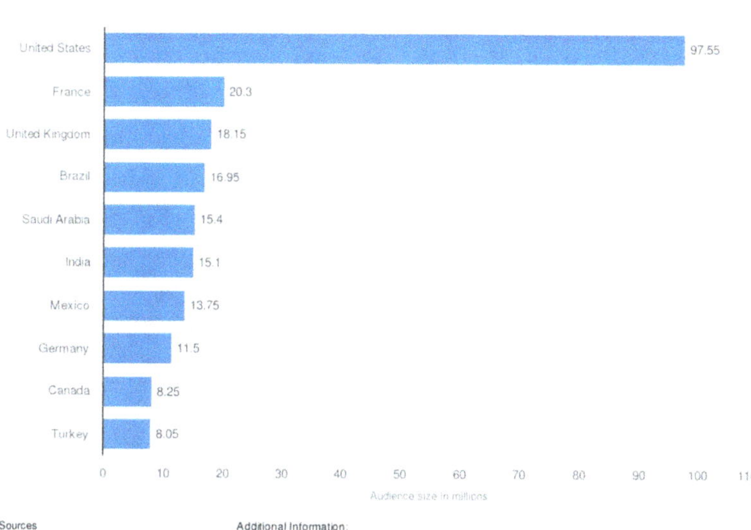

Figure 4.1: Countries with the most Snapchat users 2019

The links between mathematics and the Islamic religion, and mathematicians:

The sixth most common reasons for perceptions of why girls liked, or disliked mathematics was the links between mathematics and the Islamic religion, and Mathematicians. 5 out of 87 of the female students disliked mathematics because it is difficult. This is evidenced by quotes from one respondent. One stated, "I dislike mathematics. Mathematics is difficult because it depends on understanding, and I prefer memorising". 2 out of 87 liked mathematics because of Islamic religion. One stated, "It is useful in terms of religion. Allah said in Holy Quran 'that you might know the number of years and how to calculate time'" (Quran surah Jonah 10, P: 127; Translated by Haleem). 2 out of 87 liked mathematics because of Mathematicians, one stated " I like mathematics because of Al-Khwarizmi".

The encouragement and support received from parents:

The fifth most common reason for perceptions of why boys liked mathematics was the encouragement and support they received from parents. 8

out of 85 of the male students liked mathematics because of their parents. One respondent illustrated this finding by saying, "My father always encourages me to study mathematics", while another said, " I like mathematics because my mother is a mathematics teacher and she encourages me to study mathematics".

Several studies show that parents play a vital role in attitudes toward mathematics (Stevenson and Newman, 1986; Ma, 2001). Although social conformity is strong in Saudi Arabia, compared with countries in the West, which tend to have a more individualistic culture (Smith and Bond, 1998) , the influence of Saudi parents on their children, in terms of liking mathematics, was weak in this study. One explanation could be that the influences of parents depend on the countries' culture. For example, Cao et al. (2006) demonstrated that while parents greatly influence Chinese students' attitudes to mathematics, the same does not hold true for Australians.

Question two: an open-ended question asked girl and boy students to report on their perceptions of mathematics expressed with ''mathematics is…''. 90 out of 90 of the girls' responses to this question were analysed and 86 out of 90 of the boys' responses. 4 responses from males were excluded because they were unclear. Coding of this question resulted in two interpretable themes for both females and males. Firstly, mathematics is an arithmetic subject which helps us in daily live, science, and career. Secondly, Mathematics is addition, subtraction multiplication, division, shapes, geometry, …etc.

Mathematics is an arithmetic subject which help us in daily life, science, and career:

The most common expressions for 'what is mathematics?' was about importance of mathematics. 53 out of 90 of the female students, and 61out of 86 from males derived mathematics' importance from their beliefs that mathematics helps them in daily life and will be useful them in their careers. One female respondent illustrated this finding by saying, "Mathematics is an arithmetic subject, helps us in daily life, and it is the foundation of everything", while a another female said, " Mathematics is rules and numbers helps us in our daily life and career ". A male student said, "It is the language of numbers which helps us in our daily lives, and it is important for development", while another male said, " It solves problems of economy, and we use it to know market prices and discounts, and mathematics is useful for professional life".

Finally, based on results above, some students have an application-oriented view, demonstrating the belief that the utility of mathematics for real world problems is the main aspect of it. This view could come from four views that are called "world views" regarding mathematics, which are 'formalism', 'process', 'application', and 'scheme aspect". These views describe teachers' beliefs toward mathematics (Grigutsch et al., 1998). In the other words, students' perceptions about the importance mathematics could be due to how their teachers have affected them (Ruffell et al.,1998). (See this chapter, section 4.8, themes related to importance of mathematics and mathematics for career).

Mathematics is addition, subtraction, multiplication, division, shapes, geometry, rules, equations:

The second common expressions for 'what is mathematics?' was about these topics. 37 out of 90 of the female students, and 25 out of 86 of males expressed mathematics as addition, subtraction, multiplication, division, shapes, geometry, rules, and equations. One female respondent illustrated this finding by saying, " mathematics is addition, subtraction, multiplication, division, triangles and rules", while a one male said, " mathematics is addition, subtraction, multiplication, division, and equations".

Finally, based on results above, some students look to mathematics as abstraction subject. Abstraction, need not be a fundamental problem, especially since the data here did not support it. Most female and male students like mathematics and they did not consider mathematics to be a difficult subject. However, abstraction may be detrimental to attitudes if it deviates from reality for the learner. (See for e.g. working of Hussein (2006) in chemistry).

The summary of the two previous questions:

The results show that both female and male students liked mathematics, and their attitude appeared to be influenced by the following:

1- Usefulness of mathematics in everyday life (Both females and males).

2- The importance of the teacher (Both females and males).

3- Mathematics for career and 'future life' (Both females and males).

4- The encouragement and support received from social media personalities (Females), and parents (Males).

5- Enjoyment of mathematics (Both females and males).

6- The notable difficulty of mathematics (Both females and males).

Although, both old and recent studies indicate that girls tend to hold more negative perspectives toward mathematics in middle schools (Fennema-Sherman,1978; Ding et al, 2015), the results in this study showed female students liked mathematics more than male students. This may be due to different in cultural environment and educational context, as mentioned previously.

Chapter 5

Conclusion

Overview: After analyzing the study results in the previous chapter, this chapter summarizes the research methodology and all the research findings. It is divided into three sections. The first section summarises the methodology used to answer the research questions and the resulting findings from analysis of the data which were collected. It also gives a brief overview of the main discussion points arising from the analysis. The second section presents the research contributions, and the researcher's personal outcomes from the study. Finally, the third section discusses the limitations of the research, makes recommendations and suggests possible future work.

<h1 style="text-align:center">Chapter 5: Conclusion</h1>

5. Conclusion

5.1 Conclusion

Mathematics is an essential subject and is compulsory for middle school educational provision in most countries. It is fundamental for the study of science and for coping in everyday life. However, some students do not enjoy their studies in mathematics and cannot see any usefulness in it, leading to negative attitudes to the subject and poor exam results. This may create a vicious circle since there is a reciprocal relationship between attitude and achievement.

This study has investigated attitudes relating to mathematics, considering in particular any gender differences in the following areas:

1- Mathematics anxiety.

2- Confidence with mathematics.

3- Students' perceptions on the learning of mathematics.

4- The usefulness of mathematics.

5- Enjoyment of mathematics.

6- Mathematics class environment and the relationship with teacher and peers.

The research also considered students' perception of what mathematics is, and why they liked, or disliked mathematics.

The study employed a mixed methodology to answer the research questions. Quantitative data were collected, using a questionnaire, to answer the first research question (see chapter one) and were analysed statistically. This was supported by the addition of qualitative, open ended questions, which were analysed thematically. For the second main research question, an independent t-test was employed to assess whether there were differing average values for gender. The questionnaire was applied in two Saudi government-run middle schools (one for girls and one for boys) from different areas in Riyadh city in Saudi Arabia, giving an overall sample of 180. Students indicated their attitude towards mathematics by rating 29 statements on a 5-point scale, in the Likert scale style, coded in SPSS. The responses for each part of each statement were summarised according to gender. Open-ended questions allowed participants to provide their own responses, unguided by the researcher.

Both male and female students demonstrated positive attitudes toward mathematics. Interestingly, in all areas of this study (mathematics anxiety, confidence with mathematics, students' perceptions on the learning of mathematics, the usefulness of mathematics, enjoyment of mathematics, mathematics class environment and relationship with teacher and peers), it appears that female students tend to have more positive attitudes towards mathematics than the male students, although the weight of research evidence suggests a reversal of this trend (Dowker et al, 2016; Pajares and Graham,1999; Sherman,1978; Organisation for Economic Co-operation and Development (OECD), 2013; Frost et al., 1994; Foxman et al., 1981; Else-Questet al, 2013; Reilly et al, 2019). This is not easy to interpret, but has been attributed to the following three main reasons:

1- Cultural environment and educational context, from country to country (Cogan and Schmidt, 1999). (the same result was found by Reilly et al. (2019) in relation to Oman students)

2- The gender-role socialization pattern in Saudi culture (See Birenbaum and Nasser (2006), in relation to Jewish and Arab students in Israel).

3- Saudi educational policy, which keeps strict gender segregation. (The same result was found by Norton and Rennie (1998) in relation to Australia students, in terms of anxiety and attitudes, Streitmatter (1997) in relation to US students, in terms of confidence, Prendergast and O'Donoghue (2014) in relation to US students, in terms of enjoyment, Lee and Lockheed (1990), in relation to Nigerian students, in terms of teacher). All areas of the current study show that there are statistically significant gender differences, except area of usefulness of mathematics.

In terms of open-ended questions, the results show that both female and male students liked mathematics, and their attitude appeared to be influenced by the following:

1- Usefulness of mathematics in everyday life (both females and males).

2- The importance of the teacher (both females and males).

3- Mathematics for career and 'future life' (both females and males).

4- The encouragement and support received from social media personalities (females), and parents (males).

5- Enjoyment of mathematics (both females and males).

6- The Notable Difficulty of mathematics (both females and males).

The usefulness of mathematics emerged as the most common reasons affecting student attitudes toward mathematics in Saudi Arabia. This finding can be attributed to Saudi Arabia's new curriculum, (detailed in chapter one) which makes intensive use of examples related to daily life (McGraw-Hill, 2019), in line with the theory of Realistic Mathematics Education (RME) (Freudenthal's, 1977). The importance of the teacher emerged as the second most common reasons for perceptions of why students liked mathematics. Both female and male students' responses revealed the significance of a kind, respectful teacher and their influence in making lessons enjoyable and ensuring that students understand the work. The work of Roeser et al. (1998) supports these findings. The impact of social media personalities and internet teachers on girls' attitudes to mathematics was an important and unexpected finding. It was the third most common reasons for perceptions of why girls liked mathematics. This finding can be explained in terms of the high level of social media use among adolescent Saudi girls. This underlines the importance of teacher awareness of social media personalities, as argued by Nardi (2016). They may thus be able harness their positive impact to enhance attitudes towards mathematics amongst their students. The influence of Saudi parents on their children, in terms of liking mathematics, was weak in this study. This may relate to the findings of such researchers as Cao et al. (2006) who demonstrated that parental influence is culturally dependent. A further factor for both girls' and boys' liking of mathematics was its usefulness for their future careers, especially if they hoped to become astronomers, mathematicians or computer scientists. This is in line with Willis (2010) who stressed personal targets as a way of encouraging a liking for mathematics. A considerable level of enjoyment of mathematics was found among the participants, which is an encouraging finding given the importance of enjoyment in intrinsic motivation (Ryan and Deci, 2000). The most common reason for students to dislike mathematics was the perception that it is complicated and difficult (Sharples, 1969; Dossey et.al, 1988). However, contrary to previous studies (Foxman et al., 1982), the current study found that more boys than girls believe that mathematics is difficult.

5.2 General and Academic Contributions

In general terms, this study provides important insights into Saudi middle school students' attitudes towards mathematics and will help teachers and policy makers ensure the best ways of fostering positive attitudes towards mathematics and hence improving standards of achievement. Usefulness of mathematics in everyday life, teachers' styles, mathematics for career and 'future life', enjoyment of the mathematics and the notable difficulty of mathematics were identified as factors that influence students' attitudes towards mathematics in two Saudi government-run middle-schools. However, there is concern about the future of attitudes towards mathematics in Saudi Arabia, especially with lower achievement in TIMMS in Saudi middle school.

From an academic point of view, this research has contributed to the knowledge in this area by showing that female students tend to have more positive attitudes towards mathematics than the male students, in all areas of this study, which is unusual in comparison with other countries worldwide and suggests that important cultural and social factors are at play. Furthermore, the study has produced important new findings about the impact of social media personalities on girls' perceptions of mathematics, showing that these figures can have a strong positive or negative influential effect on female students. This is an area with great potential for practical application in schools and a promising area for future academic research.

5.3 Personal Outcomes

Conducting this research has given me many opportunities to develop my skills as an academic, a researcher and a classroom teacher. It has enhanced my understanding of students' attitudes towards mathematics in ways which will definitely inform my future approaches to lesson preparation and delivery. The research required me to read extensively and critically, in English. This has improved my language skills and my ability to make a critical assessment of research of various kinds. Furthermore, I needed to construct a research tool for this study, which led me to creating a questionnaire, based on previous work but tailored to suit the needs of my own research. This was a new and rewarding experience for me. The data analysis also gave me my first opportunity to use the SPSS. Reviewing the results of the data led me to think deeply about all the information that had been generated and make useful and critical comparison with the work of earlier researchers. This improved my thinking and my analytical skills which in turn can aid me in becoming a better thinker, communicator, and developer of ideas and research protocols. Finally, I feel that I have deepened my knowledge of this study area and developed my skills in such a way that I will be able in future to tackle the challenge of studying for a PhD.

5.4 Recommendations and Limitations

The findings of this study have useful and practical applications for the education system in Saudi Arabia. In view of the results and conclusions which have been produced, the following recommendations could be made to the Ministry of Education in Saudi Arabia:

- Given that students have a generally positive attitudes towards mathematics in this study, there are clearly important issues to address about why TIMMS results for Saudi Arabia are currently so low.
- More research needs to be conducted into why girls in Saudi Arabia have more positive attitudes towards mathematics than boys, which is contrary to most findings elsewhere in world.
- Teachers have been shown to be an important factor in student attitudes towards mathematics. It would be beneficial to introduce some in-service training to ensure that they do not teach with a view to helping students pass easy exams, but are aware of the need to teach the more challenging aspects of mathematics, in an engaging manner, so that their students will be better prepared to succeed in exams such as TIMMS. It is important for students to achieve well in such exams so that their positive attitudes are maintained for the future.
- It might be useful to conduct future studies which measure student attitudes towards mathematics on more than occasion, to see whether they develop or change in any way.
- The study has highlighted the importance of social media personalities and internet teachers in shaping girls' attitudes towards mathematics. It is therefore recommended firstly that classroom teachers are made aware of this finding and are given ways in which they can maximise the benefits of this phenomenon for their students. Secondly, it might also be useful to organize a conference on this topic and invite some of these personalities to take part, along with classroom teachers and researchers.

Any study inevitably has its limitations, but these provide opportunities to develop further and conduct more research. Firstly, this study only focused on two schools, in Riyadh and was not able to cover schools in other regions, away from the capital city. With more time and greater resources, the study could be extended to other cities and to rural areas to give a fuller picture of the country as a whole, which would make the study more generalisable.

This study used a questionnaire and two open ended questions. There is always a risk with these types of tool that the response does not fully express the views of the participants. It might therefore be useful in future to conduct interviews or focus groups to follow up the information received through the questionnaire. However, for the present researcher, it would not have been possible for me to have access to female students for interviews. It might therefore be useful in future for both a male and a female researcher to conduct the research.

Finally, the size of the study sample was comparatively small, making it more difficult to generalise the findings. It would be useful to recruit a larger sample with students from more schools across Saudi Arabia.

References:

Abdul Haleem, M. A. S. (2010). The Qur'an: English translation with parallel Arabic text.

Abrego, M. B. (1966). Children's attitudes toward arithmetic. *The Arithmetic Teacher*, *13*(3), 206-208.

Adebule, S. O., & Aborisade, O. J. (2014). Gender comparison of attitude of senior secondary school students towards mathematics in Ekiti state, Nigeria. *European Scientific Journal*, *10*(19).

Adelson, J. L., & McCoach, D. B. (2011). Development and psychometric properties of the math and me survey: Measuring third through sixth graders' attitudes toward mathematics. *Measurement and Evaluation in Counseling and Development*, *44*(4), 225-247.

Adelson, J. L., & McCoach, D. B. (2011). Development and psychometric properties of the math and me survey: Measuring third through sixth graders' attitudes toward mathematics. *Measurement and Evaluation in Counseling and Development*, *44*(4), 225-247.

Aiken Jr, L. R., & Dreger, R. M. (1961). The effect of attitudes on performance in mathematics. *Journal of Educational psychology*, *52*(1), 19.

Aiken Jr, L. R. (1976). Update on attitudes and other affective variables in learning mathematics. *Review of educational research*, *46*(2), 293-311.

Aiken, L.R. (1969). Attitudes toward mathematics. Studies in mathematics, 19, 1-49.

Allport, G. W. (1935). Attitude in CM Murchion (ed.), Handbook of social psychology.

Arab News. (2019). *Saudi men use social media more than women: Survey.* [online] Available at: https://www.arabnews.com/node/1477421/media [Accessed 29 Sep. 2019].

Asante, K. O. (2012). Secondary students' attitudes towards mathematics. *IFE Psychologia: An International Journal*, *20*(1), 121-133.

Bandura, A. (1997). *Self-efficacy: The exercise of control.* Macmillan.

Bandura, A. (1986). Social foundations of thought and action. *Englewood Cliffs, NJ, 1986.*

Beaton, A.E., Mullis, I.V.S., Martin, M.O., Gonzalez, E.J., Kelly, D.L. and Smith, T.A. (1996). *Mathematics achievement in the middle school years. Third International Mathematics and Science Study*. Center for the Study of Testing, Evaluation and Educational Policy, Boston College, MA.

Belbase, S. (2010). Images, Anxieties and Attitudes toward Mathematics. Online Submission.

Bell, E., Bryman, A., & Harley, B. (2018). *Business research methods*. Oxford university press.

Berndt, T. J. (1979). Developmental changes in conformity to peers and parents. *Developmental psychology*, *15*(6), 608.

Bill, J. M. (1973). A methodological study of the interview and questionnaire approaches to information-gathering. *Research in Education*, *9*(1), 25-42.

Birenbaum, M., & Nasser, F. (2006). Ethnic and gender differences in mathematics achievement and in dispositions towards the study of mathematics. *Learning and Instruction*, *16*(1), 26-40.

Bishop, A. J. (1988). Mathematics education in its cultural context. *Educational studies in mathematics*, *19*(2), 179-191.

Bohner, G., & Wänke, M. (2002). Attitudes and Attitude Change: Social Psychology. *Hove, UK*.

Booker, C. L., Kelly, Y. J., & Sacker, A. (2018). Gender differences in the associations between age trends of social media interaction and well-being among 10-15 year olds in the UK. *BMC public health*, *18*(1), 321.

Braun, V., & Clarke, V. (2006). Using thematic analysis in psychology. *Qualitative research in psychology*, *3*(2), 77-101.

Cao, Z., Bishop, A., & Forgasz, H. (2007). Perceived parental influence on mathematics learning: A comparison among students in China and Australia. *Educational Studies in Mathematics*, *64*(1), 85-106.

Eagly, A. H., & Chaiken, S. (1993). *The psychology of attitudes*. Harcourt Brace Jovanovich College Publishers.

Chipman, S. F. (2005). Research on the women and mathematics issues. *Gender differences in mathematics: An integrative psychological approach*, 1-24.

Çiftçi, S. K., & Yildiz, P. (2019). The Effect of Self-Confidence on Mathematics Achievement: The Metaanalysis of Trends in International Mathematics and Science Study (TIMSS). *International Journal of Instruction, 12*(2), 683-694.

Cogan, L. S., & Schmidt, W. H. (2012). An examination of instructional practices in six countries. In *International comparisons in mathematics education* (pp. 78-95). Routledge.

Cohen, L., Manion, L., Morrison, K. & Bell, R. (2011). Research methods in education. 7th edn. New York: Taylor & Francis.

Corcoran, M., & Gibb, E. G. (1961). Appraising attitudes in the learning of mathematics. *Evaluation in mathematics: Twenty-sixth yearbook of the NCTM.*

Cronbach, L. J. (1951). Coefficient alpha and the internal structure of tests. *psychometrika, 16*(3), 297-334.

Deighan, W.P. (1971). An examination of the relationship between teachers 'attitudes toward arithmetic and the attitudes of their students toward arithmetic.

Devrim, U. Z. E. L., & Uyangor, S. M. (2006). Attitudes of 7th class students toward mathematics in realistic mathematics education. In *International Mathematical Forum* (Vol. 1, No. 39, pp. 1951-1959).

Di Martino, P., & Zan, R. (2001, August). Attitude toward mathematics: some theoretical issues. In *PME CONFERENCE* (Vol. 2, pp. 3-351).

Dickinson, P., & Hough, S. (2012). Using realistic mathematics education in UK classrooms. *Centre for Mathematics Education, Manchester Metropolitan University, Manchester, UK.*

Ding, L., Pepin, B., & Jones, K. (2015). Students' attitudes towards mathematics across lower secondary schools in Shanghai. In *From beliefs to dynamic affect systems in mathematics education* (pp. 157-178). Springer, Cham.

Dobie, T. E. (2019). Expanding conceptions of utility: middle school students' perspectives on the usefulness of mathematics. *Mathematical Thinking and Learning, 21*(1), 28-53.

Dossey, J. A. (1988). *The Mathematics Report Card: Are We Measuring Up? Trends and Achievement Based on the 1986 National Assessment.* National

Assessment of Educational Progress, Educational Testing Service, Rosedale Road, Princeton, NJ 08541-0001.

Dowker, A., Sarkar, A., & Looi, C. Y. (2016). Mathematics anxiety: What have we learned in 60 years?. *Frontiers in psychology, 7*, 508.

Dreger, R. M., & Aiken Jr, L. R. (1957). The identification of number anxiety in a college population. *Journal of Educational psychology, 48*(6), 344.

Dunn, T. J., Baguley, T., & Brunsden, V. (2014). From alpha to omega: A practical solution to the pervasive problem of internal consistency estimation. *British journal of psychology, 105*(3), 399-412.

Eagly, A. H., & Chaiken, S. (1993). *The psychology of attitudes*. Harcourt Brace Jovanovich College Publishers.

Else-Quest, N. M., Mineo, C. C., & Higgins, A. (2013). Math and science attitudes and achievement at the intersection of gender and ethnicity. *Psychology of Women Quarterly, 37*(3), 293-309.

Esenhardt, W. B. (1977). A Search for Predominant Casual Sequence in the Interrelationship of Interest in Academic Subject and Academic Achievement. A crossed- lagged panel correlation study, 37.

Farooq, M. S., & Shah, S. Z. U. (2008). Students' attitude towards mathematics. *Pakistan Economic and Social Review*, 75-83.

Fennema, E., & Hart, L. E. (1994). Gender and the JRME. *Journal for Research in Mathematics Education.*

Fennema, E., & Sherman, J. A. (1976). Fennema-Sherman mathematics attitudes scales: Instruments designed to measure attitudes toward the learning of mathematics by females and males. *Journal for research in Mathematics Education, 7*(5), 324-326.

Fennema, E. H., & Sherman, J. A. (1978). Sex-related differences in mathematics achievement and related factors: A further study. *Journal for Research in Mathematics Education*, 189-203.

Field, A. (2009). Discovering Statistics Using SPSS, Sage Publications Inc. *freedom (Chicago, University of Chicago Press).*

Fishbein, M., & Ajzen, I. (1977). Belief, attitude, intention, and behavior: An introduction to theory and research.

Forgasz, H., & Leder, G. (2015). Single-sex versus co-educational schooling and STEM pathways. *Education (VCE)*, 9.

Foxman, D. D., Badger, M. E., Martini, R. M., & Mitchel (1981). Attitude Toward Mathematics Mathematics Development, Secondary Survey Report. Northern Ireland, Welsh Office.

Foxman, D. D., Martini, R. M., & Mitchel, D. (1982). Attitude Mathematics Development*: Secondary survey report* (No. 3). HM Stationery Office.

Fraser, B. J., & Fisher, D. L. (1982). Predicting students' outcomes from their perceptions of classroom psychosocial environment. *American Educational Research Journal*, *19*(4), 498-518.

Frost, L. A., Hyde, J. S., & Fennema, E. (1994). Gender, mathematics performance, and mathematics-related attitudes and affect: A meta-analytic synthesis. *International Journal of Educational Research*, *21*(4), 373-385.

Gardner, P. L. (1995). Measuring attitudes to science: Unidimensionality and internal consistency revisited. *Research in science education*, *25*(3), 283-289.

George, D., & Mallery, P. (2003). SPSS for Windows step by step: A simple guide and reference. 11.0 update (4th ed.). Boston: *Allyn & Bacon.*

Gniewosz, B., Eccles, J. S., & Noack, P. (2012). Secondary school transition and the use of different sources of information for the construction of the academic self-concept. *Social development*, *21*(3), 537-557.

Goodenow, C. (1993). The psychological sense of school membership among adolescents: Scale development and educational correlates. *Psychology in the Schools*, *30*(1), 79-90.

Gray, E. M., & Tall, D. O. (1994). Duality, ambiguity, and flexibility: A" proceptual" view of simple arithmetic. *Journal for research in Mathematics Education*, 116-140.

Grigutsch, S., Raatz, U., & Törner, G. (1998). Einstellungen gegenüber mathematik bei mathematiklehrern. *Journal für Mathematik-Didaktik*, *19*(1), 3-45.

Grimm, P. (2010). Social desirability bias. *Wiley international encyclopedia of marketing*. Hoboken: Wiley-Blackwell.

Haladyna, T., Shaughnessy, J., & Shaughnessy, J. M. (1983). A causal analysis of attitude toward mathematics. *Journal for Research in mathematics Education*, 19-29.

Hannula, M. S. (2002). Attitude towards mathematics: Emotions, expectations and values. *Educational studies in Mathematics*, *49*(1), 25-46.

Hart, L. E. (1989). Describing the affective domain: Saying what we mean. In *Affect and mathematical problem solving* (pp. 37-45). Springer, New York, NY.

Hartmut, J. (1978). Supportive dimensions of teacher behavior in relationship to pupil emotional cognitive processes. *Psychologie in Erziehung und Unterricht*, 25, 69-74.

He, H. (2007). *Adolescents' Perception of Parental and Peer Mathematics Anxiety and Attitude Toward Mathematics: A Comparative Study of European-American and Mainland-Chinese Students* (Doctoral dissertation, Washington State University).

Heise, D. R. (1970). The semantic differential and attitude research. *Attitude measurement*, *4*, 235-253.

Henerson, M. E., Morris, L. L., & Fitz-Gibbon, C. T. (1987). *How to measure attitudes*. Sage.

Hicks, L. (1997). How do academic motivation and peer relationships mix in an adolescent's world?. *Middle School Journal*, *28*(4), 18-22.

Høgheim, S., & Reber, R. (2019). Interesting, but less interested: Gender differences and similarities in mathematics interest. *Scandinavian Journal of Educational Research*, *63*(2), 285-299.

Hungerman, A. D. (1967). Achievement and attitude of sixth-grade pupils in conventional and contemporary mathematics programs. *The Arithmetic Teacher*, *14*(1), 30-39.

Hussein, F. K. A. (2006). *Exploring attitudes and difficulties in school chemistry in the Emirates* (Doctoral dissertation, University of Glasgow).

Ipsos, M. O. R. I. (2012). Four ways to get the truth out of respondents. The social research newsletter from Ipsos MORI Scotland, 1-2.

Hadden, R. A., & Johnstone, A. H. (1983). Secondary school pupils' attitudes to science: the year of erosion. *European Journal of Science Education, 5*(3), 309-318.

Johnstone, A. H., & Hadden, R. A. (1983a). Secondary School pupils' attitude to science: the year of erosion. *European Journal of Science Education, 5*, 309-318.

Johnstone, A. H., & Hadden, R. A. (1983b). Secondary school pupils' attitude to science: the year of decision. *European Journal of Science Education, 5*, 429-438.

Jung, E. S. (2005). *Attitudes and learning difficulties in middle school science in South Korea* (Doctoral dissertation, ProQuest Dissertations & Theses,).

Keller, C. (2001). Effect of teachers' stereotyping on students' stereotyping of mathematics as a male domain. *The Journal of Social Psychology, 141*(2), 165-173.

Leder, G. C., & Forgasz, H. J. (2002). Two new instruments to probe attitudes about gender and mathematics.

Lee, V. E., & Lockheed, M. E. (1990). The effects of single-sex schooling on achievement and attitudes in Nigeria. *Comparative Education Review, 34*(2), 209-231.

Lemon, N. (1973). Attitudes and their measurement. Oxford, England: John Wiley & Sons.

Liebert, R. M., & Morris, L. W. (1967). Cognitive and emotional components of test anxiety: A distinction and some initial data. *Psychological reports, 20*(3), 975-978.

Likert, R. (1932). A technique for the measurement of attitudes. *Archives of psychology.*

Locke, L.F., Spirduso, W.W. & Silverman, S.J., (2013). Proposals That Work: A Guide for Planning.

Lowry, G. & Turner, R. (2007). *Information systems and technology education. London: Idea Group Inc.*

Ma, X. (1997). Reciprocal relationships between attitude toward mathematics and achievement in mathematics. *The Journal of Educational Research, 90*(4), 221-229.

Ma, X., & Xu, J. (2004). The causal ordering of mathematics anxiety and mathematics achievement: a longitudinal panel analysis. *Journal of adolescence, 27*(2), 165-179.

Maio, G. R., Haddock, G., & Verplanken, B. (2018). *The psychology of attitudes and attitude change*. Sage Publications Limited.

Mallam, W. A. (1993). Impact of school-type and sex of the teacher on female students' attitudes toward mathematics in Nigerian secondary schools. *Educational studies in mathematics, 24*(2), 223-229.

McGraw-Hill. (2019). *McGraw-Hill*. [online] Available at: https://www.mheducation.com [Accessed 29 Sep. 2019].

McLeod, D. B. (1992). Research on affect in mathematics education: A reconceptualization. *Handbook of research on mathematics teaching and learning, 1*, 575-596.

Mendoza, Y. (1996). Developing and Implementing a Parental Awareness Program To Increase Parental Involvement and Enhance Mathematics Performance and Attitude of At-Risk Seventh Grade Students.

Ministry of Education, (2006). Development of Education in the Kingdom of Saudi Arabia. Riyadh: Ministry of Education.

Morrisett, L. M., and Vinsonhaler, J. (1965). Mathematical Learning. Monographs of the Society for Research in Child Development, 3, 132.

Muijs, D. (2004). Introduction to quantitative research. *Doing quantitative research in education with SPSS*, 1-12.

Namkung, J. M., Peng, P., & Lin, X. (2019). The Relation Between Mathematics Anxiety and Mathematics Performance Among School-Aged Students: A Meta-Analysis. *Review of Educational Research, 89*(3), 459-496.

Nardi, E. (2016). Exploring and overwriting mathematical stereotypes in the media, arts and popular culture: The visibility spectrum. khdm-Report, 5, 73-81.

Newman, R. S., & Schwager, M. T. (1993). Students' perceptions of the teacher and classmates in relation to reported help seeking in math class. *The Elementary School Journal, 94*(1), 3-17.

Norton, S. J., & Rennie, L. J. (1998). Students' attitudes towards mathematics in single-sex and coeducational schools. *Mathematics Education Research Journal, 10*(1), 16-36.

Nurmi, A., Hannula, M., Maijala, H., & Pehkonen, E. (2003). On Pupils' Self-Confidence in Mathematics: Gender Comparisons. *International Group for the Psychology of Mathematics Education, 3*, 453-460.

Oppenheim, A. N. (1992). *Questionnaire design, interviewing and attitude measurement*, London and New York.

Organisation for Economic Co-operation and Development (OECD). (2013). *PISA 2012 results: ready to learn: students' engagement, drive and self-beliefs (volume III): preliminary version*. OECD, Paris, France.

Orton, A., & Frobisher, L. (2004). *Insights into teaching mathematics*. A&C Black.

Orton, A., & Wain, G. (1994). *Issues in Teaching Mathematics*. Cassell, 387 Park Avenue South, New York, NY 10016-8810 (paperback: ISBN-0-304-32680-1; clothbound: ISBN-0-304-32678-X).

Osterman, K. F. (2000). Students' need for belonging in the school community. *Review of educational research, 70*(3), 323-367.

Pajares, F., & Graham, L. (1999). Self-efficacy, motivation constructs, and mathematics performance of entering middle school students. *Contemporary educational psychology, 24*(2), 124-139.

Pajares, F., & Miller, M. D. (1994). Role of self-efficacy and self-concept beliefs in mathematical problem solving: A path analysis. *Journal of educational psychology, 86*(2), 193.

Papanastasiou, C. (2000). Effects of Attitudes and Beliefs on Mathematics Achievement. *Studies in educational evaluation, 26*(1), 27-42.

Parahoo, K. (2014). *Nursing research: principles, process and issues*. Macmillan International Higher Education.

Parsons, S., Croft, T., & Harrison, M. (2009). Does students' confidence in their ability in mathematics matter?. *Teaching Mathematics and Its Applications: An International Journal of the IMA, 28*(2), 53-68.

Pepin, B. (2011). Pupils' attitudes towards mathematics: a comparative study of Norwegian and English secondary students. *ZDM, 43*(4), 535-546.

Popham, W. J. (2005). Students' Attitudes Count. *Educational leadership, 62*(5), 84.

Prendergast, M., & O'Donoghue, J. (2014). Influence of gender, single-sex and co-educational schooling on students' enjoyment and achievement in mathematics. *International Journal of Mathematical Education in Science and Technology, 45*(8), 1115-1130.

Ramsden, J. M. (1998). Mission impossible?: Can anything be done about attitudes to science?. *International Journal of Science Education, 20*(2), 125-137.

Recber, S., Isiksal, M., & Koç, Y. (2018). Investigating self-efficacy, anxiety, attitudes and mathematics achievement regarding gender and school type. *Anales de Psicología/Annals of Psychology, 34*(1), 41-51.

Reilly, D., Neumann, D. L., & Andrews, G. (2019). Investigating gender differences in mathematics and science: Results from the 2011 Trends in Mathematics and Science Survey. *Research in Science Education, 49*(1), 25-50.

Richardson, F. C., & Suinn, R. M. (1972). The mathematics anxiety rating scale: psychometric data. *Journal of counseling Psychology, 19*(6), 551.

Robson, J. (1996). Some outcomes of learning through teleconferencing. *Journal of Instructional Science and Technology, 1*(3).

Roeser, R. W., Eccles, J. S., & Sameroff, A. J. (1998). Academic and emotional functioning in early adolescence: Longitudinal relations, patterns, and prediction by experience in middle school. *Development and psychopathology, 10*(2), 321-352.

Ruffell, M., Mason, J., & Allen, B. (1998). Studying attitude to mathematics. *Educational Studies in Mathematics, 35*(1), 1-18.

Ryan, R. M., & Deci, E. L. (2000). Self-determination theory and the facilitation of intrinsic motivation, social development, and well-being. *American psychologist, 55*(1), 68.

Sadker, M., & Sadker, D. (1994). Failing at fairness: How our schools cheat girls (new york. *Touchstone*, 271-303.

Sait, S., Al-Tawil, K., & Hussain, S. (2004). E-Commerce in Saudi Arabia: adoption and perspectives. *Australasian Journal of Information Systems, 12*(1).

Sakiz, G., Pape, S. J., & Hoy, A. W. (2012). Does perceived teacher affective support matter for middle school students in mathematics classrooms?. *Journal of school Psychology, 50*(2), 235-255.

Shapiro, E. W. (1961). *Attitudes toward arithmetic among public school children in the intermediate grades* (Doctoral dissertation, University of Denver).

Sharples, D. (1969). Children's attitudes towards junior school activities. *British Journal of Educational Psychology, 39*(1), 72-77.

Skemp, R. R. (1976). Relational understanding and instrumental understanding. *Mathematics teaching, 77*(1), 20-26.

Smith, P. B., & Bond, M. H. (1993). Social Psychology Across Cultures: Analysis and Perspectives (Harvester Wheatsheaf, Hemel Hempstead).

Soleymani, B., & Rekabdar, G. (2016). Relation between math self-efficacy and mathematics achievement with control of math attitude. *Applied Mathematics, 6*(1), 16-19.

Star, J. R., & Stylianides, G. J. (2013). Procedural and conceptual knowledge: exploring the gap between knowledge type and knowledge quality. *Canadian Journal of Science, Mathematics and Technology Education, 13*(2), 169-181.

Statista. (2019). *Countries with most Snapchat users 2019 | Statista.* [online] Available at: https://www.statista.com/statistics/315405/snapchat-user-region-distribution [Accessed 29 Sep. 2019].

Stromquist, N. P. (2000). On truth, voice, and qualitative research. *International Journal of Qualitative Studies in Education, 13*(2), 139-152.

Syyeda, F. (2016). Understanding Attitudes Towards Mathematics (ATM) using a Multimodal Model: An Exploratory Case Study with Secondary School Children in England. Cambridge Open-Review Educational Research e-Journal, 3, 32-62.

Tahar, N. F., Ismail, Z., Zamani, N. D., & Adnan, N. (2010). Students' attitude toward mathematics: The use of factor analysis in determining the criteria. *Procedia-Social and Behavioral Sciences, 8*, 476-481.

Teo, T. (Ed.). (2014). *Handbook of quantitative methods for educational research.* Springer Science & Business Media.

Thurstone, L. L., & Chave, E. J. (1929). The measurement of attitude: Chicago, University of Chicago Press.

Thurstone, L., L (1929). Attitude Can be Measured. Psychological Review, 36, 222- 241.

Thurstone, L. L. (1931). The measurement of social attitudes. *The journal of abnormal and social psychology*, 26(3), 249.

Achievement, S. (2019). *Multiple Comparisons of Average Mathematics Achievement – TIMSS 2015 and TIMSS Advanced 2015 International Results*. [online] Timss2015.org. Available at: http://timss2015.org/timss-2015/mathematics/student-achievement/multiple-comparisons-of-mathematics-achievement/ [Accessed 29 Sep. 2019].

Torrance, E. (1966). Characteristics of mathematics teachers that affect students' learning.

Uwineza, I., Rubagiza, J., Hakizimana, T., & Uwamahoro, J. (2018). Gender attitudes and perceptions towards mathematics performance and enrolment in Rwandan secondary schools. *Rwandan Journal of Education*, 4(2), 44-56.

Vision2030. (2019). *رؤية المملكة العربية السعودية 2030*. [online] Available at: http://vision2030.gov.sa/ [Accessed 29 Sep. 2019].

Walonick, D. S. (1993). Everything you wanted to know about questionnaires but were afraid to ask.

Wigfield, A., & Meece, J. L. (1988). Math anxiety in elementary and secondary school students. *Journal of educational Psychology*, 80(2), 210.

Wiles, C. A. (1992). Investigating Gender Bias in the Evaluations of Middle School Teachers of Mathematics. *School Science and Mathematics*, 92(6), 295-98.

Willis, J. (2010). *Learning to love math: teaching strategies that change student attitudes and get results*. ASCD.

Wong, K. Y., & Chen, Q. (2012). *Nature of an attitudes toward learning mathematics questionnaire*.

Yang, X. (2013). Investigation of junior secondary students' perceptions of mathematics classroom learning environments in China. *Eurasia Journal of Mathematics, Science & Technology Education*, 9(3), 273-284.

Yara, P. O. (2009). Relationship between teachers' attitude and students' academic achievement in mathematics in some selected senior secondary schools in southwestern Nigeria. *European Journal of Social Sciences*, 11(3), 364-369.

Yin, R. K. (1984). Case study research: design and methods (Beverley Hills, CA, Sage).

Zimmerman, B. J. (2000). Self-efficacy: An essential motive to learn. *Contemporary educational psychology*, 25(1), 82-91.

Appendices:

Appendix A: Consent from girls' school

[Translation into English]

Dear Mr. Majed Saeed Alharthi

May the peace and blessings of Allah be upon you

In accordance to your e-mail sent on (25/07/1440 AH) regarding your request of obtaining a permission to conduct a study in gathering data from the Middle School of 52nd for females in the Riyadh, we advise that we are granting the permission of gathering data throughout your sister Mrs. Fauzia Saeed Alharthi from the period between 25/07/1440 AH to 16/08/1440 AH. Therefore, this letter of permission been granted upon your request to be sent to the Saudi Arabian Cultural Bureau Mission in the United Kingdom.

Wishing you all the best for your dissertation.

School Principle,

Lamia Al-Bilaji

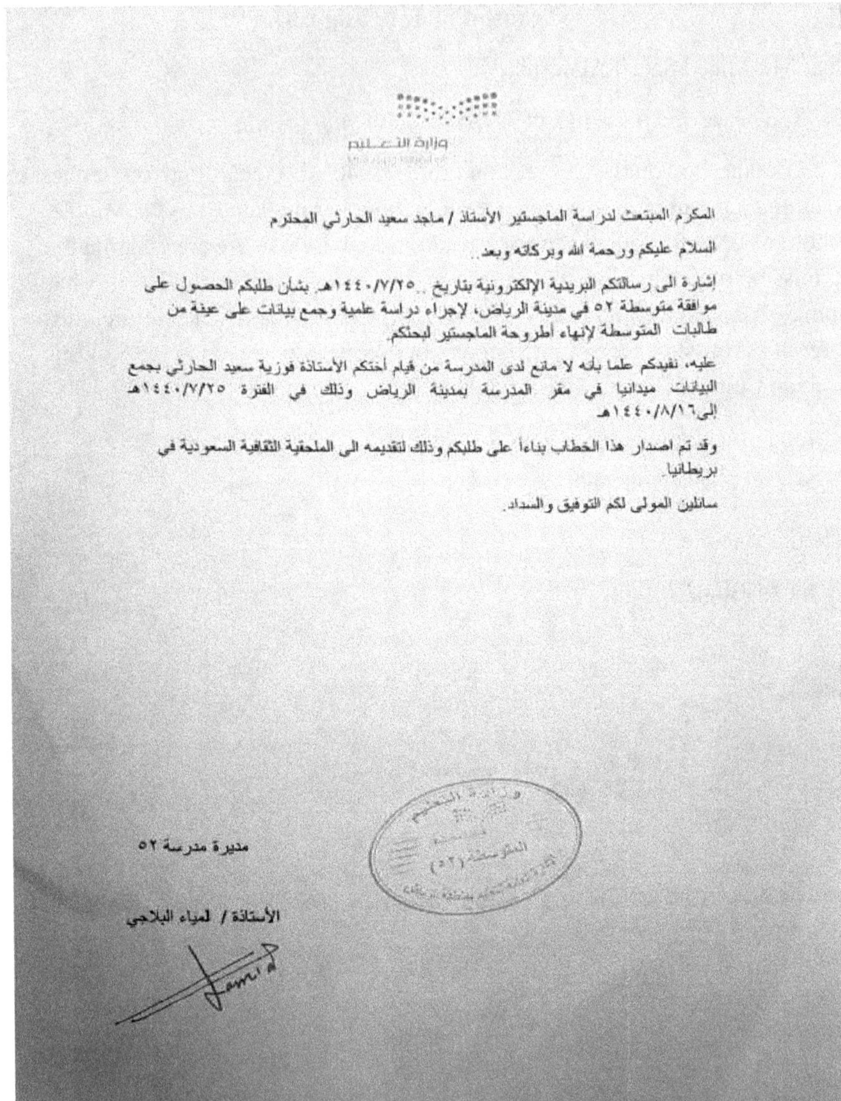

المكرم المبتعث لدراسة الماجستير الأستاذ / ماجد سعيد الحارثي المحترم

السلام عليكم ورحمة الله وبركاته وبعد ..

إشارة الى رسالتكم البريدية الإلكترونية بتاريخ ١٤٤٠/٧/٢٥هـ بشأن طلبكم الحصول على موافقة متوسطة ٥٢ في مدينة الرياض، لإجراء دراسة علمية وجمع بيانات على عينة من طالبات المتوسطة لإنهاء أطروحة الماجستير لبحثكم.

عليه، نفيدكم علما بأنه لا مانع لدى المدرسة من قيام أختكم الأستاذة فوزية سعيد الحارثي بجمع البيانات ميدانيا في مقر المدرسة بمدينة الرياض وذلك في الفترة ١٤٤٠/٧/٢٥هـ إلى ١٤٤٠/٨/١٦هـ

وقد تم اصدار هذا الخطاب بناءً على طلبكم وذلك لتقديمه الى الملحقية الثقافية السعودية في بريطانيا.

سائلين المولى لكم التوفيق والسداد.

مديرة مدرسة ٥٢

الأستاذة / لمياء البلاجي

Appendix B: Consent from boys' school

[Translation into English]

Dear Mr. Majed Saeed Alharthi

May the peace and blessings of Allah be upon you

In accordance to your e-mail sent on (25/07/1440 AH) regarding your request of obtaining a permission to conduct a study in gathering data from the Middle School of Al-Imam As-Sousi for males in the Riyadh, we advise that we are granting the permission of gathering data to yourself from the period between 25/07/1440 AH to 16/08/1440 AH. Therefore, this letter of permission been granted upon your request to be sent to the Saudi Arabian Cultural Bureau Mission in the United Kingdom.

Wishing you all the best for your dissertation.

Mohammed Abdullah Al-Yahya

المكرم المبتعث لدراسة الماجستير الأستاذ/ ماجد بن سعيد الحارثي   المحترم

السلام عليكم ورحمة الله وبركاته .... وبعد:

إشارة الى رسالتكم الالكترونية بتاريخ ١٤٤٠/٠٧/٢٥هـ بشأن طلبكم الحصول على موافقة متوسطة الامام السوسي لتحفيظ القرآن الكريم في مدينة الرياض، لأجراء دراسة علمية وجمع البيانات عينية من طلاب المتوسطة لإنهاء أطروحة الماجستير لبحثكم.

عليه نفيدكم علمًا بانه لا مانع لدى المدرسة من قيامكم وذلك للفترة من ١٤٤٠/٠٧/٢٥هـ الى ١٤٤٠/٠٨/١٦هـ بجمع البيانات ميدانيًا في مقر المدرسة بمدينة الرياض.

وقد تم اصدار هذا الخطاب بناءً على طلبكم وذلك لتقديمه الى الملحقية الثقافية السعودية في بريطانيا.

سائلين المولى لكم التوفيق والسداد،،،،

محمد بن عبد الله اليحيى

83

Appendix C: Participant information statement
Consent from the headmasters of school (English)

Dear Principal,

I am writing to you about my research project that I am conducting as part of my Master study in School of Education and Lifelong Learning at the University of East Anglia in the United Kingdom. "My research has title "Investigation of students' attitudes toward mathematics in middle schools in Saudi Arabia?". The overall aim of this research is to explore attitude toward mathematics in Saudi middle schools.". In addition, this research will suggest some solutions to improve students' attitude toward mathematics. The study uses questionnaire to collect data which will be subjected to quantitative analysis.

You can read the information sheet about the questionnaire. Your agreement for student/s in your school to take part in this study is going to be a factor in the success of this research. If you have a willingness to help me in carrying out this study; I hope you to sign this consent and return it to me in the attached envelope. If you have any additional questions or would like further information about this study; please contact me at M.Alharthi@uea.ac.uk or my supervisor Prof Irene Biza - I.Biza@uea.ac.uk

Yours sincerely,
The Researcher/ Majed Alharthi
School of Education and Lifelong Learning
University of East Anglia
NORWICH NR4 7TJ
The United Kingdom
M.Alharthi@uea.ac.uk

The principal agreement
I agree ....
I do not agree ....
Signature...........................

Consent from the family member (English)

Dear family,

I am writing to you about my research project that I am conducting as part of my Master study in School of Education and Lifelong Learning at the University of East Anglia in the United Kingdom. "My research has title "Investigation of students' attitudes toward mathematics in middle schools in Saudi Arabia?". The overall aim of this research is to explore attitude toward mathematics in Saudi middle schools.". In addition, this research will suggest some solutions to improve students' attitude toward mathematics. Moreover, I will explore "Is there any significant gender difference in students' attitudes towards mathematics?" The study uses questionnaire to collect data which will be subjected to quantitative analysis. As it is known, The Ministry of Education in Saudi Arabia follow a policy of gender – based separation at all levels. Therefore, I need you to help me in carrying out this study. You can read the information sheet about the questionnaire. Your agreement for student/s in your school to take part in this study is going to be a factor in the success of this research. I hope you to sign this consent and return it to me in the attached envelope. If you have any additional questions or would like further information about this study; please contact me at M.Alharthi@uea.ac.uk or my supervisor Prof Irene Biza - I.Biza@uea.ac.uk

Yours sincerely,

The Researcher/ Majed Alharthi

School of Education and Lifelong Learning

University of East Anglia

NORWICH NR4 7TJ

The United Kingdom

M.Alharthi@uea.ac.uk

The family member agreement

I agree ………………………………

I do not agree ………………………

Signature……………………………

MAJED ALHARTHI

MA Education

26/02/2019

**Faculty of Social
Sciences**

School of Education and
Lifelong Learning
University of East
Anglia

**In which ways can improve students' attitude toward
mathematics in middle school in Saudi?**

PARENTAL INFORMATION STATEMENT

(1) What is this study about?

Your child is invited to take part in a research study titled ""Investigation of
students' attitudes toward mathematics in middle schools in Saudi Arabia?".
The overall aim of this research is to explore attitude toward mathematics in
Saudi middle schools.". In addition, this research will suggest some solutions to
improve students' attitude toward mathematics. The study uses questionnaire
to collect data which will be subjected to quantitative analysis. We are
interested in how your child thinks and feels about the mathematics and Your
child's views will help me to make an assessment of the current situation and
enable me to make recommendations to educators and decision-makers in
Saudi Arabia to improve students' attitude toward mathematics. This
Participant Information Statement tells you about the research study. Knowing
what is involved will help you decide if you want to let your child take part in
the research. Please read this sheet carefully and ask questions about anything
that you don't understand or want to know more about. Participation in this
research study is voluntary. By giving your consent you are telling us that you:

- ✓   Understand what you have read.
- ✓   Agree for your child to take part in the research study as outlined below.
- ✓   Agree to the use of your child's personal information as described.
- ✓   You have received a copy of this Parental Information Statement to keep.

(2) Who is running the study?

The study is being carried out by the following student, MAJED ALHARTHI, who
is conducting this study as the basis for the degree of MA Education:

Mathematics Education. at the University of East Anglia. This will take place under the supervision of Prof Irene Biza.

(3)   What will the study involve?
Your child will be asked to complete an anonymous paper questionnaire that will ask your child questions about his/her attitude toward mathematics.

(4)   How much of my child's time will the study take?
It is expected that the survey will take between 10-15 minutes to complete.

(5)   Does my child have to be in the study? Can they withdraw from the study once they've started?
Being in this study is completely voluntary and your child does not have to take part. Your decision whether to let them participate will not affect your/their relationship with the researchers or anyone else at the University of East Anglia. If you decide to let your child take part in the study and then change your mind later (or they no longer wish to take part), they are free to withdraw from the study at any time. Your child's questionnaire responses can be withdrawn any time before they have submitted the questionnaire. Once they have submitted it, their responses cannot be withdrawn because they are anonymous and therefore we will not be able to tell which one is theirs.

(6)   Are there any risks or costs associated with being in the study?
Aside from giving up their time, we do not expect that there will be any risks or costs associated with taking part in this study for your child.

(7)   Are there any benefits associated with being in the study?
Your child responses are likely to provide details about attitude towards mathematics. In addition, your child's views will help me to make an assessment of the current situation and enable me to make recommendations to educators and decision-makers in Saudi Arabia to improve students' attitude toward mathematics.

(8) What will happen to information that is collected during the study?

By providing your consent, you are agreeing to us collecting personal information about your child for the purposes of this research study. Their information will only be used for the purposes outlined in this Participant Information Statement, unless you consent otherwise. Data management will follow the 2018 General Data Protection Regulation Act and the University of East Anglia Research Data Management Policy (2015). Your child's information will be stored securely, and their identity/information will be kept strictly confidential, except as required by law. Study findings may be published, but your child will not be identified in these publications if you decide to let your child take part in the research. In this instance, data will be stored for a period of 10 years and then destroyed.

(9) What if we would like further information about the study?

When you have read this information, MAJED ALHARTHI will be available to discuss it with you further and answer any questions you may have. If you would like to know more at any stage during the study, please contact my supervisor Prof Irene Biza - I.Biza@uea.ac.uk.

(10) Will I be told the results of the study?

You and your child have a right to receive feedback about the overall results of this study. You can tell us that you wish to receive feedback by ticking the relevant box on the consent form. This feedback will be in the form of a one-page summary and you will receive this feedback after the study is finished.

(11) What if we have a complaint or any concerns about the study?

The ethical aspects of this study have been approved under the regulations of the University of East Anglia's School of Education and Lifelong Learning Research Ethics Committee.If there is a problem please let me know. You can contact me via the University at the following address:

Majed Alharthi

School of Education and Lifelong Learning

University of East Anglia

NORWICH NR4 7TJ

The United Kingdom

M.Alharthi@uea.ac.uk

If you would like to speak to someone else you can contact my supervisor: Prof Irene Biza - I.Biza@uea.ac.uk

If you or your child are concerned about the way this study is being conducted or you wish to make a complaint to someone independent from the study, please contact the Head of the School of Education and Lifelong Learning, Professor Richard Andrews at Richard.Andrews@uea.ac.uk.

(12) OK, I'm happy for my child to take part – what do I do next?
If you are happy and consent to your child taking part in the study simply allow them to complete the questionnaire provided and answer the questions. By allowing your child to submit their responses you are agreeing to the researcher using the data collected for the purposes described above. You need to fill in one copy of the consent form and ask your child to return this to me . Please keep the information sheet for your information.

This information sheet is for you to keep

PARENT/CARER CONSENT FORM (1st Copy to Researcher)

I, ……………………………………………………………… [PRINT PARENT'S/CARER'S NAME], consent to my child ………………………………………………………….[PRINT CHILD'S NAME] participating in this research study.

In giving my consent I state that:

✓ I understand the purpose of the study, what my child will be asked to do, and any risks/benefits involved.

✓ I have read the Information Statement and have been able to discuss my child's involvement in the study with the researchers if I wished to do so.

✓ The researchers have answered any questions that I had about the study and I am happy with the answers.

✓ I understand that being in this study is completely voluntary and my child does not have to take part. My decision whether to let them take part in the study will not affect our relationship with the researchers or anyone else at the University of East Anglia now or in the future.I understand that my child can withdraw from the study at any time.

✓ I understand that my child is completing an anonymous questionnaire and once they have submitted the questionnaire they will not be able to withdraw their data.

✓ I understand that personal information about my child that is collected over the course of this project will be stored securely and will only be used for purposes that I have agreed to. I understand that information about my child will only be told to others with my permission, except as required by law.

✓ I understand that the results of this study will be used for a dissertation assessment but that it will not contain my child's name or any identifiable information about them.

I consent to:

PARENT/CARER CONSENT FORM (2nd Copy to Parent/Carer)

I, ………………………………………………………………………… [PRINT PARENT'S/CARER'S NAME], consent to my child …………………………………………………………………….[PRINT CHILD'S NAME] participating in this research study.

In giving my consent I state that:

✓ I understand the purpose of the study, what my child will be asked to do, and any risks/benefits involved.

✓ I have read the Information Statement and have been able to discuss my child's involvement in the study with the researchers if I wished to do so.

✓ The researchers have answered any questions that I had about the study and I am happy with the answers.

✓ I understand that being in this study is completely voluntary and my child does not have to take part. My decision whether to let them take part in the study will not affect our relationship with the researchers or anyone else at the University of East Anglia now or in the future.I understand that my child can withdraw from the study at any time.

✓ I understand that my child is completing an anonymous questionnaire and once they have submitted the questionnaire they will not be able to withdraw their data.

✓ I understand that personal information about my child that is collected over the course of this project will be stored securely and will only be used for purposes that I have agreed to. I understand that information about my child will only be told to others with my permission, except as required by law.

✓ I understand that the results of this study will be used for a dissertation assessment but that it will not contain my child's name or any identifiable information about them.

I consent to:

Appendix D: Data Analysis

Independent Samples Test

		Levene's Test for Equality of Variances		t-test for Equality of Means						95% Confidence Interval of the Difference	
		F	Sig.	t	df	Sig. (2-tailed)	Mean Difference	Std. Error Difference	Lower	Upper
Mathematics Anxiety	Equal variances assumed	5.190	.024	-3.655	178	.000	-.48889	.13376	-.75286	-.22492
	Equal variances not assumed			-3.655	169.431	.000	-.48889	.13376	-.75295	-.22483
Confidence with mathematics	Equal variances assumed	4.652	.032	-2.743	178	.007	-.24259	.08844	-.41712	-.06807
	Equal variances not assumed			-2.743	173.443	.007	-.24259	.08844	-.41715	-.06804
Self-Perception	Equal variances assumed	.132	.716	.576	178	.565	.07778	.13507	-.18877	.34433
	Equal variances not assumed			.576	177.582	.565	.07778	.13507	-.18878	.34433
Perceived Usefulness	Equal variances assumed	2.564	.111	-1.824	178	.070	-.19365	.10617	-.40316	.01586
	Equal variances not assumed			-1.824	175.263	.070	-.19365	.10617	-.40318	.01588
Enjoyment of Mathematics	Equal variances assumed	.956	.329	-3.864	178	.000	-.53333	.13803	-.80572	-.26095
	Equal variances not assumed			-3.864	176.149	.000	-.53333	.13803	-.80574	-.26093
Social Factors	Equal variances assumed	1.357	.246	-3.249	178	.001	-.27963	.08606	-.44947	-.10979
	Equal variances not assumed			-3.249	177.984	.001	-.27963	.08606	-.44947	-.10979
attitude towards mathematics	Equal variances assumed	2.253	.135	-3.746	178	.000	-.28889	.07712	-.44108	-.13670
	Equal variances not assumed			-3.746	175.501	.000	-.28889	.07712	-.44110	-.13668

Group Statistics

	Gender	N	Mean	Std. Deviation	Std. Error Mean
Mathematics Anxiety	Female	90	2.3511	.79000	.08327
	Male	90	2.8400	.99310	.10468
Confidence with mathematics	Female	90	1.8389	.54306	.05724
	Male	90	2.0815	.63954	.06741
Self-Perception	Female	90	3.6167	.88384	.09317
	Male	90	3.5389	.92781	.09780
Perceived Usefulness	Female	90	1.8254	.66621	.07022
	Male	90	2.0190	.75538	.07962
Enjoyment of Mathematics	Female	90	1.8481	.87718	.09246
	Male	90	2.3815	.97222	.10248
Social Factors	Female	90	1.8407	.57459	.06057
	Male	90	2.1204	.58006	.06114
attitude towards mathematics	Female	90	2.0479	.48551	.05118
	Male	90	2.3368	.54735	.05770